Humans Are Not Resources

Gez Smith

Odrilea Ltd
2020

Dedicated to Jemima and Ember, my two lovely ladies

Release Notes

Version	Date	Notes
0.9	September 2019	Initial review copy of book to gain customer feedback. May still contain typos, and require additional proof reading for flow.
1.0	June 2020	First Edition

Contents

Acknowledgements

As ever, a book is in reality the collected efforts of a number of different people, so it's only right that we acknowledge their input.

First, I have to thank Rick Barker, my co-trainer, mentor and general brother-from-another-mother, with whom I've spent so much of the last three years travelling the world whilst training, coaching, thinking and reflecting. I wouldn't have had so many of the experiences that have led to ideas for this book if it hadn't have been for him.

Next, the people in my wider agile journey who have inspired me and pointed my learning in interesting directions. Most notably Ryan and Meghan at the Scrum Alliance who inspired me to write the first few blog posts that then escalated somewhat into this, a whole book of ideas. Likewise all of the people at Scaled Agile Inc, especially Dean Leffingwell, Drew Jemilo, Joe Vallone, Steve Mayner, Andrew Sales, Erin Humbach and Bria Schecker, who have supported and encouraged my learning into all sorts of new areas, and given me an audience to test these ideas out on too.

My reviewers too, the people who took the time out to read the original version of this book, and give me such comprehensive feedback and powerful inspiration. Most notably Simon Chesney, who inspired a fantastic pivot in this book's direction, and Steph South, whose forensic analysis of the book from an HR perspective was priceless.

Finally, my lovely partner Bev, who sat patiently on long summer evenings whilst I typed away furiously, and put up so well with the occasional "*Yep, hang on, I won't be a minute, I just need to scribble down this idea whilst I remember it…*".

Thank you all. I'm hugely and humbly grateful for your support

Foreword

I don't mean to write books you know, they just sort of happen.

Each book I write though is born out of a unique set of circumstances, and this book is no different. I've been scribbling down ideas and blog posts in the lean-agile people management and HR space for a few years now, because I really do believe that it's one of the biggest areas that still needs to be tackled if the promises of lean and agile are to come true and revolutionise the modern workplace.

It's all very well telling people they're free, autonomous and collectively accountable, but if policies, procedures and hierarchies are still set up for a world where people are managed, controlled and individually assessed, all whilst being measured on hours worked rather than value delivered, then lean and agile are pretty undermined right from the start. For the power that management holds over people's careers, families and lives is immense, and its requirements are hard to ignore.

The thing is though, this is still a very new field. Lean and agile have both been around for many years now, but only relatively recently have they started to spread into the sorts of large corporations where extensive hierarchies and management using heavily codified policies and procedures are so prevalent. On top of this, the fundamental point of lean and agile is about creating an organisation that learns. One that continuously inspects reality as it exists, not as we wish it were, and then uses that empirical evidence to learn how to make things better.

It is for this reason that this book is not going to tell you exactly what to do. The field of agile management is too new, and your own specific contexts will all be too different anyway. Instead, this is a book of ideas, each one with a hypothesis, a thing I believe to be true, and an experiment you can run in your workplace to decide for yourself whether it is true or not. As such, you could read this book all in one go if you wanted, but equally I'd like to think you could dip in and out of it over time, trying out different experiments and seeing what happens.

Some ideas may appeal to you more than others, which is entirely as it should be. Some may seem completely unfeasible or ridiculous in your organisational context, and that's completely fine too. I'm not here to tell you what to do. I just hope to point you in various interesting directions so you can see where you end up yourself.

All I would ask is that if you do try out some of these ideas and experiments, please do tell me the results. I'd love to know how they do or do not work for you, and I'd ideally like to use them to improve future versions of this book. So we can all learn together as a virtual community of lean-agile people management practitioners. The knowledge you gain from your experiments could be shared to make your whole profession a better place.

If you want to get in touch with your findings, thoughts or just general feedback, my email address is gez@odrilea.com.

It would be genuinely lovely to hear from you.

Gez Smith
Glastonbury, June 2020

Section 1. Introduction

First things first, this is not a book that is going to teach you how to adopt lean and agile in how you deliver your own work day to day. This is not to say that you shouldn't do so. In fact, as we'll look at in idea number 1, I'd strongly recommend that you do. However, there are already lots of books and courses out there that can get you started with the basics of adopting lean and agile in how you deliver your own work. It didn't seem worth writing yet another book covering that same old ground.

Instead of looking at *how* you should deliver your work, this book takes a more holistic view, to look at *what* work people managers will need to deliver in order to support their wider organisation in realising the benefits of lean and agile. In essence, regardless of whether they deliver their work using agile or not, what work will people managers need to deliver to support their organisation's lean and agile efforts? It is this question that this book aims to answer.

However, I'm conscious that there will be people reading this book who have no idea what lean and agile are, which might make some of the ideas in this book tricky to understand. So it seems sensible to spend some time exploring that, so you have some context for what the rest of this book contains.

Let's start with lean. Lean is an approach to delivering work that some say can be traced back to the late 1800's in Japan, but it was more widely formalized and popularized in Japan after the Second World War, as the Japanese manufacturing sector started to blossom and boom. It is especially well known thanks to the working methods developed by the Toyota car production company, which were then written down and published in a book called The Toyota Way (Liker, 2004).

There are a lot of different ideas in lean, but a lot of it is about creating a culture of continuous improvement in an organisation, through relentlessly looking at what is happening and identifying and removing any waste from the whole delivery system.

It does this through practices such as making the work that people are doing visible, measuring and levelling out its flow through the system, focusing on quality and continuously growing and developing the people who do the work. As such, it is as much a philosophy and belief system as it is a set of recommended practices and processes. In fact, there's a comment in The Toyota Way book that notes how Toyota were unconcerned about American managers coming to Japan to see how they worked, as whilst these managers may have been able to copy Toyota's processes and procedures, unless they also truly understood and embraced the lean mindset that underpinned it all, then they were largely wasting their time.

Agile as an approach appeared slightly later than lean, although there are claims that you can see the origins of agile approaches in some military equipment development in the 1960's. Primarily though, the approaches that went on to become known as agile began in the late 1980's and early 1990's, and it's important to note how they started.

During the late 80's and early 90's, various different people were experimenting with finding better ways for teams to work, especially in the field of software development. They were beginning to notice that traditional project management delivery methods were struggling, even failing, when applied to areas like software development. These traditional approaches broadly claimed that you could design and specify up front what work needed to be done, then estimate and plan out how long it would all take and what it would cost. Then, with all that locked down, you could do the work by following the plan, test it all to check it works as specified, then release it.

What became apparent was that this may work in simple, predictable environments, such a building a bridge, where you can work out up front how much concrete you need, and be confident that both the concrete and the gravity acting on it will behave in predictable ways.

However, this 'big up front design' approach starts to fall apart when work becomes more complex and unpredictable.

For example, software development was a relatively new field. It turned out to be hard to predict up front what the customer really wanted, and how the work to deliver it would progress. Even if you could have done these things, you still couldn't know how your competitors and the external market were going to behave whilst you delivered this big long piece of work. The work was complex, fast-paced, uncertain and constantly changing, and so was the external market environment for which it was being created. It was difficult to say at the start what needed to be built, how long it would take to build and how much it would cost. Nor could you know what would change around you as you completed the work.

As a result, various people and teams began to experiment with new ways of working that could better handle this sort of complexity and uncertainty. These new approaches involved the customer much more closely and much more frequently throughout the delivery process. They built rapid prototypes to get real feedback based on working systems, rather than documenting large specifications on paper. They only planned a small amount of work, then delivered it, then got feedback on what had been done, in order to use the feedback to help plan the next small increment of delivery.

The people trying out these ideas started to write down what did and did not work from the experiments they ran, and their writings became known as frameworks. So for example, there was the Scrum framework created by Jeff Sutherland and Ken Schwaber, or the Extreme Programming (XP) approach developed by Kent Beck, Ron Jeffries and Ward Cunningham. In 2001, Alistair Cockburn brought together 17 of the people working on these different sorts of approaches in a ski resort in Snowbird, Utah, USA, to see if they could find a common thread and cause to unite their efforts. Over the course of a few days, they wrote 'The Manifesto For Agile Software Development', and so the word 'agile' in the context of this book was born.

For all that this manifesto is the closest we've really got to setting out exactly 'what agile is', it has to be said that as a document it doesn't really tell you what to do. It sets out four value statements and 12 principles. The value statements state;

"We are uncovering better ways of developing software by doing it and helping others do it. Through this work we have come to value:

Individuals and interactions over processes and tools
Working software over comprehensive documentation
Customer collaboration over contract negotiation
Responding to change over following a plan

That is, while there is value in the items on the right,
we value the items on the left more." [1]

Which is great, but it's not exactly an action plan for agile adoption. There are four sets of two contrasting things, and each of them is valuable, only one of them is to some degree more valuable than the other, but we're not going to say by how much. The principles are similar, and we'll pick out a number of them as we go through this book, but again they don't tell you exactly what to do. Instead, they give you broad principles to check your work against.

In a way, this is reflective of how the manifesto was created. It brought together people who were creating different frameworks that did tell you what to do, and tried to unite them using less prescriptive values and principles that were broad enough to satisfy everyone. This is therefore the best way to think about agile. It is a set of values and principles, sometimes described as a 'mindset', to which you can align yourself in how you approach your work, but they're not going to give you exact processes and procedures to follow.

---

[1] http://www.agilemanifesto.org

If you do want to adopt an agile approach, then you probably need to start adopting one of the frameworks that now exist in the agile space, such as Scrum or XP. Having said that, there are many that will disagree with me about that, and to some degree they've got a point.

Just like lean, agile is also about continuous improvement, constantly inspecting and adapting what you deliver and how you deliver it, in order to improve both of these things. So even if you do follow a framework, doing so won't do you any good unless you are also open to changing and improving how you work; unless you are open to adopting the mindset that sits behind it.

For example, in the Scrum framework we find things like the 'Daily Scrum' (sometimes called a 'standup'), a team meeting held whilst standing up at the start of the day. In the meeting, the team gets together to inspect what it did yesterday, and to plan how it is going to work together today, in order to improve continuously.

This is probably one of the best known agile 'processes', but if you miss the essential point that the meeting is about inspecting and adapting what you do and how you do it, then you end up just standing up and giving a bland status report to those around you, before going back to work by yourself for the rest of the day. Just like with the American managers visiting Toyota, unless you have a mindset of continuous experimentation and improvement, just copying the processes in an agile framework won't do you much good at all.

It is for this reason that some people in the agile community can be a bit dismissive of frameworks, as they end up being used as just processes to follow, not tools you use to foster continuous experimentation, learning and improvement. However, as much as developing this agile continuous improvement culture and mindset is essential, I tend to believe that the best way to change mindset and culture is to change your practices. For example, if you meet the same people in the same ways every day, your culture won't change. If you start meeting different people in different ways (such as in a standup), then your culture may well start to shift. So I'd say start with a framework, start working differently, then use those things as starting points to build on, in order to find your own best way of becoming agile over time

It's probably also worth at this point reflecting on the current state of agile as you may find it out there in the marketplace and on the Internet. Agile has become big news and big business over the last decade, especially as digital disruption has started to cause large corporations to become worried about how they can compete with small, nimble start up companies. As a result, a lot of money has flooded into the sector, and where there's a lot of money, a lot of people come running after it. Consequently, there are now lots of people out there calling themselves 'agile coaches', many with little to no agile experience, some of whom are just traditional project managers who have rebadged themselves. Never forget that currently you need zero qualifications or experience to call yourself an 'agile coach'.

With this influx of poorly experienced people has come an influx of poorly written blog posts, Linkedin content and so much else besides. To the point that it can be hard to understand what agile is from reading the debates and arguments online.

However, this influx of people has not necessarily been all bad news. In recent years, new people have also brought new ideas in from outside of the original agile space, especially when it comes to using agile approaches in larger organisations. For example, both the Large Scale Scrum (LeSS) framework and the Scaled Agile Framework (SAFe) have brought with them approaches from the field of systems thinking. Systems thinking is a truly fascinating area that explains so much about why large organisations struggle with their own size and complexity. It encourages us to look at the whole system and how the different parts of it interact, rather than just looking at the work we do or the specific area for which our team is responsible[2]. Agile is all the richer for all of these new voices competing for attention within it.

So as you go through this book, remember that it has been written with the spirit of lean and agile at its heart. It won't tell you what to do, you have to work that out for yourself. It will though give you ideas and experiments to try, so you can have some useful starting points for your own journey of continuous improvement.

---

[2] If you want a great introduction to systems thinking, I *hugely* recommend reading The Fifth Discipline by Peter Senge.

Section 2: Managers Adopting Lean And Agile

Idea 1. Adopt Lean And Agile Practices Yourself

> **Hypothesis**: Some people learn new practices and mindsets by directly experiencing them.
>
> **Experiment**: Try out your own small, limited experiments with lean and agile processes to better understand how lean and agile teams like to work.

There are two directions this book could have taken. One was the direction it has taken, explaining many of the different things people people managers and HR professionals may need to stop doing, start doing or just do differently in order to help their wider organisation adopt lean and agile ways of working. This is a direction I've not really seen anyone else explore, and there are certainly enough ideas to try out, combined with a significant need for them to happen.

The other way though is the way you may perhaps have been expecting. A book setting out how managers can become more lean and agile themselves in their day to day work. Perhaps you've been told you need to 'adopt the agile mindset', as if it's that easy to unpick and rebuild an entire career, perhaps an entire lifetime, of ways of thinking and working. I'd love to say it is, but the human brain doesn't work like that. It's going to take time, diligence and practice to really change the way you think and experience the world, perhaps many years.

This is one of the reasons this book doesn't really cover how people managers should discover how to 'adopt the agile mindset' themselves, because there are already so many books and courses out there to teach you how to become more lean and agile. Adding another one to that pile may not have added much value, and I doubt a book can teach you that much in isolation anyway. However, this is not to say there is no value in the topic itself, quite the opposite in fact.

As you probably know, different people learn in different ways. Some people, and I include myself in this category, learn through reading and thinking, alone in a quiet room. Some people learn through being taught amongst other people in a more classroom based setting. Others only learn by directly doing and experiencing something. My long time co-trainer, mentor and inspiration Rick Barker is one of these people. He can happily read books, and take interesting notes from them, but he himself admits the best way for him to learn something is to actually start to practice it in reality.

As a result, I'd definitely recommend that any manager wanting to support a lean-agile transformation starts to adopt some lean and agile practices themselves, in order to experience directly what feels like for the people they will be supporting in their new world. However, I deliberately say adopt *some* practices, as for all this book isn't going to teach you how to be lean and agile yourself, I think there's a common mistake people make in this area that I'd like to flag up to you.

The problem comes from the fact that most people embark on this journey from a very 'process and methodology' background and mindset. The prevalence of this deep-rooted belief system is vast, and it stems largely from the hugely influential work of a man called Frederick Winslow Taylor, and his book 'Principles of Scientific Management', written over 100 years ago in 1911. For Taylor, gaining greater efficiency and productivity in the workplace was all about looking at the processes people followed, and trying to optimise them in order to discover 'one best way'. A term that lives on to this day in the idea of 'best practice' that people often seem so keen to uncover.

The problem is that this approach is very different from what lean and agile are recommending. Sure, both lean and agile have approaches and general rules within them that work more often than not, and there are some good initial starting points for people wanting to move into these areas. However, the fundamental point of lean and agile lies in continuous experimentation and learning, in 'inspecting and adapting' as the Scrum framework puts it. You try something out based on what you're experiencing at the time, and see if it works. If it does work, great, keep doing it. If it doesn't, great, you've learned something and you can use that learning to inform your next experiment.

As a result, there's no one best way any more, and not really much best practice. There's just the current way you're trying out, and the next thing you're going to try doing in order to make things that little bit better. You're constantly looking to experiment, to learn, to improve and to change. That's what matters.

The problem that people often run into when they try to adopt lean and agile practices is that they miss this fact, because they've been brought up to believe that success is all about following a process or methodology. As a result, they book themselves and their fellow team members on a course to learn how to adopt the Scrum framework for example, and then they come back to the office and start following it. They get a whiteboard and a load of sticky notes, they stand up around it together every morning and say what they worked on yesterday, what they're going to work on today, and mention anything that's blocking their progress. Then they sit down again and carry on with their working day exactly the same as they always have done. They're following the process as faithfully as they can, but nothing really seems to be getting better. So eventually they decide this whole agile thing is just the way work had always been, or is 'just common sense' and drift back to how they always worked before.

The thing they're missing is that the tools and techniques in lean and agile are just that, tools and techniques. It's how you use them to make things different and make things better that matters. So if I were you, and I was just starting out in this area, I wouldn't stress too much about getting all of the tools and techniques perfect right from the start. I certainly wouldn't worry about 'mindset change', as that generally only changes once you've been using different tools and techniques for a long time. Instead, I'd focus far more on how to ensure that genuine team learning and team improvement can happen. Spend time on getting people comfortable with the idea that it's ok to experiment, and that it's ok for that experiment to fail, as long as you learn something from it and fail better next time round. The tools and techniques in lean and agile can help you do this for sure, but you could start to try just a few of them out and see what happens, spending your time on building a culture of experimentation and learning, rather than spending all your time trying to learn them all, and hoping that by following the new process, everything will become magically better. All tools and techniques can ever do is show you where your problems are. You have to want to fix them yourselves.

For example, a simple experiment could be to start limiting your personal Work-In-Progress, sometimes referred to as WIP, basically the number of things you're working on at any one time. Make your work visible to everyone else in your team using a whiteboard and sticky notes, then see how much work is distributed across different work statuses, such as 'To Do', 'In Progress' and 'Done', then see if by working on fewer things at once, by limiting your WIP, you could actually start finishing things rather than constantly starting new work that you never get to finish.

If you go 'too far' and set such a low limit on your WIP that suddenly you're sitting there doing nothing, take a look at whether you could start collaborating more with other people to get their most important work over the line together. Simple small experiments like these are likely to be far more useful when it comes to learning about lean and agile over time, rather than trying to dive into everything head first and getting lost amongst a sea of new ideas.

So yes, there's huge value in people managers learning about and practicing lean and agile ideas, and I'd recommend all of them do so if they want to gain a whole new perspective on what these ideas will mean in reality for the people the manage and support. However, be very careful not to adopt the new practices as if they were just a new process to follow, then wonder why nothing has changed, or why you don't feel like you have a new 'mindset'. Start small, start simple, and inspect, adapt and learn as you go. After all, the longest journey starts with a single step.

Idea 2. Inspect And Adapt Your Privacy Culture

Hypothesis: An innate culture of privacy may cause people managers to struggle to understand or adapt to areas of a lean-agile transformation, hindering its success.

Experiment: Run a desk-based exercise, grouping people management issues or areas of work into those that must be kept private, and those which may now benefit from being made transparent, with a bias towards transparency by default.

We need to talk about the culture within people management practices, especially HR departments. Now culture is a complex thing, that varies from person to person and organisation to organisation, but there's one cultural aspect I've noticed in every people manager and HR department I've worked with, and that's a culture of privacy.

I suspect this culture comes from the sort of work that these professionals often do, and in some senses it's no bad thing. When managing people, you're very often dealing with personal and interpersonal issues. You find out things about people that they would consider to be intensely private and personal. The decisions you make can extend way beyond the workplace and into people's personal lives and private prosperity. As a result, you treat all of that information with the privacy and confidentiality it deserves.

This is entirely right and proper. However, often this approach moves from being a set of practices to deal with specific situations, and becomes the all-pervading culture applied to everything related to people management that managers come into contact with. With such a regular need to keep things private, and such a risk being present if this privacy goes wrong, privacy becomes the default stance taken towards everything a people manager does.

In traditional situations of course, this isn't much of a problem. People work within their specialisms and silos, and no one really focuses on knowing what other people are doing too much. If anything, non-people management professionals are glad to see their colleagues maintaining such a culture of privacy and confidentiality. However, when you move into lean-agile environments, it can become a real problem.

Most agile frameworks are founded on the idea of transparency. The idea that you make everything visible, so you can inspect it and work out how to adapt or improve it. So people management professionals working with agile teams will implicitly be expected to match that culture of transparency. Only this then becomes a massive clash with their existing culture of privacy. At best this causes tension, at worst this can cause real harm for lean and agile teams, when people managers are busily pushing agile anti-patterns back into teams, telling them what they can and cannot be transparent about.

It's also really hard when people managers are working within an agile framework to deliver something alongside agile teams too. Agile needs short feedback loops in order to respond to change, but how do you get feedback on work when your default position is to keep everything private?

I experienced this some years ago in a conversation with an HR person who was developing a new approach to organisational training. The work they were talking about was great, so I suggested that they conduct a regular sprint review or demo around it to get regular feedback from the teams the training would be being rolled out to. They looked both confused and horrified. It was *their* job to decide what training people needed, it was the teams' jobs to receive the training at the appropriate time, and there was no way the HR person was going to tell the teams what might be going on ahead of time. A situation that sounds odd written down, but when I dug a little deeper, it became clear that it came straight out of their culture of privacy.

So how do we deal with this problem? Now I'm not going to claim that people management should become a 100% transparent practice, with everyone's personal issues being published openly on the company intranet as standard, of course not. However, what I have found useful is to get people managers who are working in lean-agile delivery of some sort to stop and consciously reflect on what parts of their work actually require privacy, and which might instead benefit from bringing in transparency. Ideally the default option should be to be transparent unless a need for privacy can be proven, but to start with, it's often just useful to open the conversation up and see where it goes.

People managers have a huge amount to bring to the success of a lean-agile transformation, and I think a huge amount to gain from adopting lean-agile approaches themselves. However, to do so, this inherent tension between the cultures of privacy and transparency will need to be resolved.

Idea 3. Make Policies Explicit

Hypothesis: People can make better use of policies and procedures when they are clearly codified and visible.
Experiment: Bring together all of your people management policies and procedures into one simple place, and make them widely available for anyone to read.

Around 450 BC, Ancient Rome had a problem. There was a long running sense of conflict between the patricians and the plebeians, that is to say the ruling families and the ordinary citizens. To settle this conflict, both sides started working on agreeing the laws that should govern the Romans, given that before this, the laws were more about social convention and common understanding rather than anything written down and agreed. After going to see what other countries did, and holding public debates, they settled on something they called the 'Twelve Tables'; a set of rights and duties that applied to each and every Roman citizen.

Now, having a codified and agreed set of laws, rights, policies or procedures is nothing especially remarkable. However, it is what they then did with the Twelve Tables that was quite interesting. They engraved them on sheets of brass, and stuck them up in a public location for everyone to read. If you wanted to know what your rights and responsibilities were, or wanted to settle an argument around them, all you had to do was walk over to the Twelve Tables and see what the law said.

Interestingly, this approach is now found in the lean idea of kanban. In his book, 'Kanban – *Successful Evolutionary Change for your Technology Business*', David Anderson talks about the different elements you need for kanban to work, and one of them is 'making policies explicit'. In this sense, policies can mean things like the criteria a piece of work needs to meet in order to move to the next workflow state, but the idea behind it is the same. If you want a system to work well, you need to agree with everyone what the rules of the system are, and then make them visible so everyone can see them. Just like the Romans did with the Twelve Tables.

Now compare these two examples with your current people management policies and procedures. I'm sure that most, probably all of them are written down somewhere. There's too much law involved in the world of employment these days for them not to be written down. However, if an employee wanted to find them, where would they start? Are they all held in one clear location within your Intranet, or does their structure reflect their evolution, in that they're spread across multiple different Intranet pages, that have been added to and amended each time a change has been agreed? If someone wants to find an answer to an HR issue, is it easier just to go and ask an HR colleague than to try to look it up themselves?

The irony is that if you're in this situation, you're experiencing exactly the same problem that many large organisations have been experiencing with their software delivery. Having grown and evolved over time, their software systems have become complex, bloated and difficult to maintain. Often people have taken parts of the code out of the software system and worked on them for a long period of time before trying to integrate them back into the system, causing all sorts of problems around version control and integration, as the system has been changed in unexpected ways since the code was taken out. In addition to this problem of integration, it is also really hard to give feedback on the system, as it's not easy to see what it does and how it works until it's fully integrated.

My suspicion is that people management policies in some organisations are in a similar situation. Parts of them have been taken out to be updated over time, causing the overall policy approach to fragment, leading to poor integration and subsequent confusion over what is and isn't policy. Even if this isn't the case, there are still likely to be issues around transparency and ease of access, making it easier just to ask someone rather than try to self-serve in the realm of HR. It also prevents something we talk about in idea 4, allowing people to give continuous feedback on people management policies and procedures. The irony is of course that HR people would very often prefer people to self-serve, especially on the simpler or more routine issues, freeing up their time to work on the tougher problems and interventions.

So, this doesn't have to be a big piece of work, but it could hold large amounts of value for your organisation. Gather together your people management policies and procedures from across the Intranet. Integrate them into a simple, plain language document and put them up somewhere clearly visible for all to see, just like the ancient Romans did with the Twelve Tables.

Idea 4. Close The Feedback Loop On People Management Policies

Hypothesis: Ownership of people management policies and procedures is sometimes centralised within a single department, with large, slow feedback loops on its work, when policies could benefit from being more rapidly responsive and decentralized.

Experiment: Tap into feedback being generated all the time by agile teams, use it to inform your people management policies and procedures, and even see which elements of this work could be decentralised down to the teams themselves.

This idea might seem obvious for anyone who has been working in the lean and agile space for any length of time, but it may not be one that has occurred yet to as many people in the people management space. The fundamental principle is this. Lean and agile are all about continuous learning and improvement, 'inspecting and adapting' as they call it in the Scrum framework. With that in mind, how often do you use feedback to inspect and adapt the people management work within your organisation?

Now it would be foolish of me to claim that you don't, and I suspect the reality is that many people management professionals already do incorporate feedback into their work. Operating as they do at the intersections between organisational strategy, real people and very often the law, people management is a profession that has to consult and get feedback on its work in order to find the best balance between these sometimes-competing interests. However, after many years on the receiving end of corporate policies and procedures, my suspicion is that lean and agile approaches could add a huge amount more to the feedback and engagement people management professionals already seek to gain around their work.

For example, when formulating new policies, how is feedback taken on them? Very likely, the formulation of the policy is itself run as a project, using traditional project management approaches, with tollgates, review stages and formal sign off procedures. There is feedback gained through these mechanisms for sure, but is it not a form of feedback that is slow, heavy weight and difficult to digest?

What if new policies and procedures were developed using more lean and agile approaches? Testing multiple different options quickly by working on them in short iterations of time, opening them out widely for feedback using public demo and review sessions, then discarding the least viable options before iterating some more on the options that survive, until the right option is left standing.

Once the policy or procedure has been set live, it could remain under constant review, open to tweaking and improvement based on real-life feedback from the people it affects, so that it in effect becomes a product that is constantly improved, rather than an output that is produced once and then left unchanged. Or at least until the pressure of change or obsolescence becomes just too overwhelming, by which time some damage may have been done.

On top of this, how are ideas for new policies and procedures created in the first place? Presumably a close eye is kept on both the legal environment the organisation is operating within, as well as the wider corporate strategy. But what if data were regularly being taken from across the different teams within an organisation, looking at their issues and patterns of behaviour, then seeing if people management interventions could help? Most agile teams have some manner of retrospective process baked into their approach, regularly stopping to look at their problems and how to solve them, so that feedback data is just sitting there waiting for people management professionals to tap into it and help the teams and people work even better.

I cover it more in the next idea, idea five, but when it comes to lean and agile in people management, there's also the idea of a shift in how

HR professionals operate, from being a centralised, specialist function that sets the policies and procedures people must follow, to becoming a more supporting and coaching function that enables others to take on people management work themselves.

For example, if agile teams are meant to be self-organising and self-managing, surely they need to gain some degree of understanding of people management policies and procedures as part of their collective self-management skillset? This shift would take HR professionals more out amongst the teams on the ground, which places them nicely in what lean calls the 'Gemba'; the real place, the place where the work happens, and the place where you can learn the most about the reality of what is going on. This provides access to another hugely important source of feedback for people management policies; direct visibility of the realities on the ground, along with the suggestions from the people on the ground as to how to improve things.

Finally, there's another angle to this feedback loop. The benefit from feedback loops is increased when they are kept small and short. When you can try something quickly, observe the results it creates, then use that learning to decide what to do next. This is at the heart of what being agile means, the ability to learn and change direction quickly. As a result, there likely has to be a trade off here between the knowledge and skills experienced people managers bring to an issue, against the inevitable slight delay and increased overhead that involving more people will bring to a problem.

Consequently, might there be some instances where the best feedback loops for people managers would be for them not to be involved in an area at all, and to hand off responsibility for that area to the teams themselves? Take for example the idea of a capability framework covered more extensively in idea 29. Often these frameworks are created at too abstract a level away from the teams to make any sense or be of much use to people on the ground. Due to their high level and slightly monolithic nature, they often end up rarely changing, even though the realities of the skills and competencies needed on the ground might be evolving rapidly.

As a result, why not devolve ownership of capability frameworks down to teams themselves, so that they can be the ones to run the experiments, gather feedback and make adjustments directly, more closely linked to their specific local context? Of course some degree of oversight would likely still be needed over this, but it could be much more supportive and light touch, perhaps aggregating the results of the decisions and experiments run in the teams, and looking for successful patterns that could be shared with others?

I hope the above has given you some ideas for how the feedback loops within the field of people management policies and procedures could be made both shorter and more effective, whilst simultaneously empowering people and teams even further in order to facilitate their moves towards lean and agile approaches. As I say, I doubt very much that you are not taking feedback on your work, but perhaps it's worth a moment's reflection on the techniques you are using to do so, and whether there are opportunities for inspecting and adapting them too.

Idea 5. Make People Management Skills Truly Cross-Functional

Hypothesis: Agile teams should be cross-functional. That is to say, they need to possess the skills and experience needed to tackle issues that emerge in the team. Some of these issues will be people management related.

Experiment: People managers should move their role from being a centralised one to which people turn for advice, to a coaching and training one that shares their knowledge far and wide, so that everyone in a team has a basic level of people management related competency.

One of the fundamental elements of an agile framework like Scrum is that people in teams should become ever more cross-functional. That is to say, they should become ever more able to undertake the roles of those around them. For example, the Scrum framework only has three roles in it; Product Owner, Scrum Master and Development Team Member. The Development Team Member role title is important, for as the Scrum Guide says;

"Individual Development Team members may have specialized skills and areas of focus, but accountability belongs to the Development Team as a whole." [3]

The role title signifies that everyone on the team is collectively accountable for what the team delivers, as they all have the same title. So if accountability for the team's results is collective, then it makes sense that everyone on the team should have some level of competency and understanding of the roles of everyone else on the team. Not only will this assist with team collaboration and communication, but it acts as an insurance policy. For if a key member of the team suddenly goes

---

[3] http://www.scrumguides.org

off sick, what does it matter, if the rest of the team can cover their absence well enough until they return? It also helps everyone on the team undertake better quality work too. If a developer understands what a tester is testing for, they are more likely to write better and more easily testable code. If a designer knows how a developer works, they can better create designs that can be implemented successfully.

So often though, the idea of knowledge and skill sharing amongst team members to enable them to become ever more cross-functional ends up stopping at the boundary of the team. But what if there are skills from outside of the team that the team members would benefit from learning? What if team members would benefit from being taught people management skills too? Let me give you two examples I've encountered where this might have been of use.

The first happened a few years ago, when I was invited in by a Scrum Master to attend a team's retrospective meeting. The meeting where the team meets up to look at how things have gone over the last iteration of work, in this case their last two weeks, and talk about any problems they've encountered and how to overcome them.

What is discussed at the meeting is up to the team itself to own and control, so the Scrum Master started out by handing everyone some sticky notes and pens, then asking them to write down any problems they'd like to talk about from the last two weeks, and stick them on the wall. The idea was that people would then vote on which problems they thought were the most important, and work through them in turn, from the most popular to the least popular.

However, as each team member wrote things down and stuck them to the wall, something horrifying became apparent. Each team member was only writing one thing on one note, and every single one of them had written the same thing. The name of a person who wasn't in the team. Then each of them stuck this person's name to the wall. The Scrum Master and I sat there for a moment, staring at each other with slightly panicked expressions, then said "Well, it looks like we've got an

HR issue here then doesn't it?".

The only problem was, we had no idea how to deal with it. Were we allowed to hold a retrospective to discuss problems with a colleague who wasn't in the room to give their side of the story or defend themselves? Could they rightfully launch a grievance procedure if we did? Would anything said in the room need to be written down in case it got used at an employment tribunal? We had no idea, but the implications of what was happening in the room seemed serious and real. If only we'd had some professional knowledge to know how to handle that situation for the best outcome. If only we'd understood and used the relevant people management skills earlier on in this process, to ensure successful mediation between people that would have prevented them from being escalated to a retrospective. If only a people management expert had taught us how to deal with issues like this professionally and legally when they arose.

The second example I'd give to support the idea of people managers transferring their knowledge and skills to team members more generally is perhaps a happier one. The Scrum Master in Scrum is meant to be a servant leader for the team. They deal with the problems that get in the team's way. They help protect the team from unhelpful interruptions, and they do whatever else the team needs them to do. In other words, they serve the team.

Yet too often, Scrum Masters for teams are interviewed by recruiters and chosen by senior people above the team. If the Scrum Master is there to serve the team though, shouldn't they be interviewed and chosen by the team themselves? Of course they should, and some great agile organisations very much make sure that this happens. But how do people like software developers learn the skills needed to carry out fair and effective interviews with candidates? Well, experienced people managers need to teach them.

I'm sure there are many other examples that can be found where the skills and experience of people managers could add huge value to lean-agile teams if opportunities were created for skills to be shared. This may though require a fundamental recasting of the role of the people manager, from being experts to whom people take problems, to experts who teach others how to be experts too.

Section 3: Resourcing

Idea 6. Tell Them They Don't Need More Headcount

> **Hypothesis**: Requests for more headcount to speed up delivery will often in reality only slow delivery down.
> **Experiment**: Only approve headcount requests incrementally, approving each new request based on empirical proof that the previous one achieved its stated goal.

How often do you hear the complaint from team managers that they don't have enough people to get the job done?

Most likely they're under huge pressure to deliver. After all, senior leaders adopted this whole lean-agile thing to ensure that your organisation could keep up with its competitors, and these team managers just don't feel like they're being given the resources to deliver everything that's now being expected of them.

Sounds familiar? So, tough as it is, you do what you can to find them more headcount, right?

Well, whilst this may seem like the obvious thing to do, and it's clearly the thing they're asking you for, the truth is that you might actually be doing them more harm than good. You see, the belief that more headcount leads to more delivery is actually mistaken, at least when it comes to agile.

For over 100 years now, whether we realise it or not, we've been brought up to believe in the production line system of delivery. One person does their work on whatever it is that is being produced, then it moves down the line to the next person, who does their work, and so on until the thing is delivered.

There's nothing wrong with this system. It works fantastically well for delivering physical products, where the work to produce them is simple, well understood and repeatable. So if you want to produce more products, you can simply add more people to the production line, and you will deliver more. One production line produces one car per hour, so five production lines will produce five cars per hour. It's a pretty indisputable fact, so no wonder it's become the dominant mental model. More people equals more work.

However, when it comes to delivering software, or any sort of creative, non-physical knowledge based work, work that is not simple, well understood and repeatable, this model completely breaks down. Indeed, what happens is often the opposite of what you think will happen. Software development for example is complex work that requires intellectual rather than physical labour. It is work that requires constant testing and exploration, regularly doing small amounts of planning and delivery to get feedback on what to build next and how to build it. That's basically what an agile framework like Scrum is.

The reason that the traditional production line model starts to break down in this situation is relatively simple. On the production line, people don't need to speak to each other. They operate their machine, and that's pretty much all they do. They pull a lever, a widget is produced, and that widget goes off to the next stage on the production line to have some more work done on it.

In software development, people need to collaborate and communicate with each other. It's complex, uncertain work that requires thinking, communication and collaboration. So for every new person you add into the software development process, you add in a new person that everyone else needs to collaborate and communicate with. If you look at the math of this, every new person added (n) adds n-1 new lines of communication. So if you've a team of 6, there are 15 different lines of communication between them. Add in one new person, and you have 21 different lines of communication. Add in one other person, and you now have 28 lines of communication.

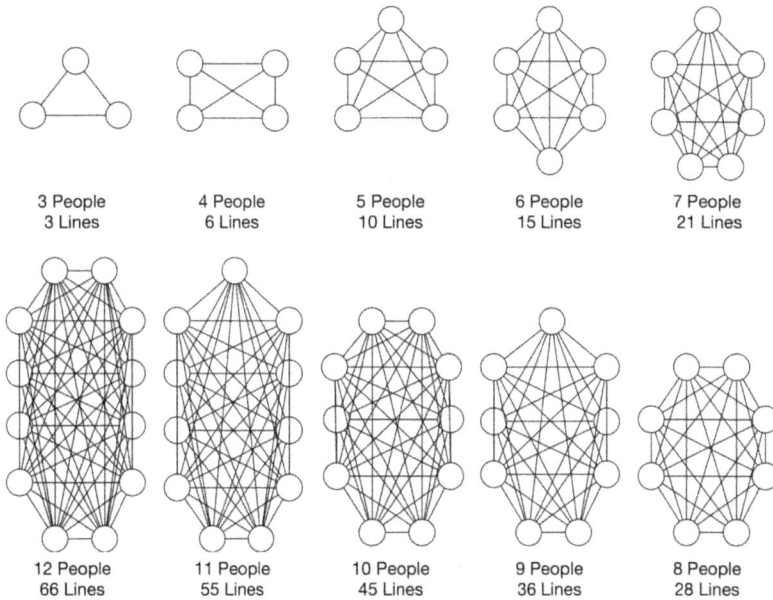

Fig. 1. Number of people in a team against number of potential lines of communication between those people.

So in a world of software development, where you're already trying to manage complexity and uncertainty, the more people you add, the more complexity you create, and the more time you spend managing that complexity, rather than delivering the work. This is one of the reasons you get Brookes' Law, which states;

> *"Adding more people to an already late software development project makes it later."*

This is the opposite of the production line of course, where adding more people and more machines leads to greater output. Where work is complex and uncertain, adding more people makes things worse not better. An increase in complexity and a slowing down of delivery is not the only problem that hiring more people creates though. It also creates self-reinforcing downward spirals. For example, the delivery starts to run late, so the manager asks for more people. This makes the project run later, which makes the manager ask for more people. Which makes the project run later, which means…you get the picture.

So next time someone comes to you asking for more people, bare Brookes' Law in mind. It may well be that by finding these people additional headcount, you're actually making their delivery later, and your organisation's problems worse.

Idea 7. Focus On Value Delivery, Not Attendance

> **Hypothesis**: Tracking the number of hours people work is a hangover from our school days that becomes a meaningless, even counter-productive, metric in the workplace.
> **Experiment**: Stop tracking the hours people work, and instead track the value they deliver, regardless of when and how much they work.

Our school days leave us with many funny quirks and ingrained beliefs, some we may not even be aware of. After all, if we're brought up from a young age to believe that something is true, then why would we even give it any more thought, let alone question it if we did happen to think about it?

I firmly believe one of these quirks is the obsession we have around the number of hours people are present at work. You see, in countries where education is mandated by law up until a certain age, they need a way of assessing who is complying with the law and who isn't. They can't really do this by looking at who passes the exams and who fails them, as there could be all sorts of reasons why people do or do not do well in exams, and you'd probably drive down the standard of exams if you did set this as a metric anyway. What you can measure though is attendance. Are all of the children present at school for all the hours of all of the days they are meant to be?

The first job I got after my first masters degree was running a research project into school attendance in a large UK city, and it struck me even then that people were obsessed with whether or not the children were present in school, or authorised to be absent if they were not present. People seemed less concerned about whether children were actually happy and learning things whilst at school, and even less interested in the idea that there might be a link between a child's happiness and engagement in learning and their lack of attendance in the first place. All that mattered was that they were recorded as present in school every day.

As a result, I believe that the idea of 100% attendance has spilled from the school into the workplace in quite a powerful yet unthinking way. Again, presumably because it provides a simple and measurable metric, our work contracts set out how many hours we are expected to be at work, and we can be penalised if we don't meet this target. Indeed, it becomes such a powerful target, that people often pride themselves on regularly exceeding it. Like some personal badge of pride, they're sure to drop casually into conversation just how many times they've worked late, or worked over their weekends and holidays, regardless of the impact this will be having on their health and personal relationships.

The problem is that whilst attendance is a primary metric for our school days, because it would not be fair to judge people on their exam results instead, the same is not true for the workplace. Whilst we all have to attend school and generally follow a standardised curriculum (Math, Science and History are common across many countries' education systems, Epistemology, Pattern Welding and Horticulture are not), we don't all have to do the same job of work.

Indeed, you'd hope that many of us could choose the field we work in due to our interests or natural abilities, and decide to change fields if we found that our job did not suit us. Thus, whilst it may not be fair to judge children on the outcomes of their education, as they have no choice over it, surely it is fair to judge adults on the outcomes of their employment, given they will hopefully have had some degree of choice over their role?

So why then are we still judging people on attendance at work when a far more useful metric is available to judge them by? That metric being the value they deliver to their organisation. Do organisations not care most of all about the value that their employees create for them? Why should someone be praised for working more hours than they are contracted to do, without asking the more important question of what value have they actually created by working those extra hours?

Once you start thinking along these lines, you have to ask, why do hours present at work matter at all? What would happen if people worked fewer hours, but still delivered the same, or even more value?

This is not as ridiculous an idea as it sounds. More and more, organisations are finding that people can deliver the same or more value if they work for less time, and their home and personal lives benefit significantly as a result too. This brings many benefits to the person's general health and wellbeing, meaning they can work better in the time they are at work. It's not by accident that principle eight of the Agile Manifesto states;

> *"Agile processes promote sustainable development.*
> *The sponsors, developers, and users should be able to*
> *maintain a constant pace indefinitely."* [4]

How better to be able to maintain a constant pace than by reducing the hours you work and spending more time with the ones you love?

I can think of many different times in my career I've been subject to these two different approaches, either of time management or value delivery, and the ones where I have been most able to be successful were the ones where my manager focused on value delivery rather than the number of hours I worked.

For example, when it came to time management, as a teenager I once had a role doing data entry, manually typing the contents of paper forms into a database, and one day I went to go to lunch at 12:59pm. I'd just finished the batch I was working on, and it seemed like a natural break point. As I headed towards the door, my manager pulled me up on it, stopping to lecture me on not leaving before 1pm, as I was "stealing time from the company". So what happened when I came

---

[4] http://www.agilemanifesto.org

back from lunch? I worked far slower than I had before, my motivation to help the company pretty much evaporated, and I delivered far less value.

In contrast, when I worked as a product owner at a small digital start-up, I don't think hours were mentioned anywhere in our contracts at all, other than we weren't allowed to start work before 10am, because none of us were morning people. As a result, I focused instead on value delivery, and delivered some of the best work I'd ever done up until that point, sometimes doing the work at odd hours of the day and night, and sometimes taking random mornings and afternoons off to compensate. I've no idea how many hours I worked in that job, because it didn't matter. I do know though that I delivered tons and tons of value to the company and our customers.

Alongside this of course comes the issue of remote working. The idea that everyone needs to be physically present in an office for a set amount of time goes hand in hand with the ideas drummed into us at school. But again, as long as we're delivering the value we need to, what does it matter if we're present in the office, or working at home, or working from somewhere completely different? I've got a friend who's a quite brilliant software developer, and she often works from ski resorts around Europe, as she snowboards, then writes code, then goes snowboarding again. Some of her clients have no idea she's doing this, but what does it matter as long as they're getting what they wanted, which they always do?

Shifting from assessing people on the number of hours they are present in an office to the amount of value they deliver, regardless of when or where they do it, will be quite a shift for many people and many organisations. After all, presenteeism has been drummed into us from school age onwards. Making the shift though will be a hugely important part of creating a lean-agile culture within an organisation, one that demonstrates clearly that its primary focus is value delivery, and that it trusts people to get the job done, no matter when or where they choose to do it.

Idea 8. Should Lean-Agile Teams Be 100% Utilised?

> **Hypothesis**: Your organisation has an implicit culture that focuses on tracking percentage time utilisation of each employee.
> **Experiment**: Start to track the value delivery of teams instead, and see how it correlates against their self-reported percentage utilisation, in order to see if reducing utilisation might help to increase value delivery.

As we looked at in the previous idea, from our earliest days at school, we're taught one thing; attendance matters. You have to be at school all day every day, or have a very good reason not to be. In some countries, if you don't attend school regularly, your parents could even be prosecuted for your non-attendance as a criminal offence.

Not only that, but when you go to school, you don't get to sit there doing nothing. Especially during your more senior school years, you get given a timetable at the start of the year, which makes sure that every hour of your day at school has a clear purpose. It's someone's job to make sure you're there every day, and also to make sure you're 100% utilised whilst you're there.

Is it any surprise then that this mindset carries on into your work life? Just like schools, companies have systems of authorised and unauthorised absence, with sanctions for unauthorised absence. Some companies I've seen even have people employed to make sure everyone is 100% utilised all time, people sometimes called 'resource managers' or similar. At more senior levels, executives have personal assistants whose job, as one exec once said to me, "is to keep their diary 100% full all of the time".

The thing is, what is it that actually generates revenue and profit for your organisation? Is it this 100% resource utilisation? Does the customer pay you in return for keeping everyone 100% busy? No, they pay you in return for delivering them some value. So in and of itself 100% resource utilisation isn't something that generates revenue and profit. At best, it can only be something that leads to value delivery, which then leads to revenue and profit.

However, not only is 100% utilisation not inherently valuable, the problem is that very often, it actually slows down the delivery of value to customers, leading to less revenue and profit. There are various reasons for this.

If people are 100% utilised, they have zero slack in their system to respond to change and the emergence of new work. They can't take on anything new, as they have literally no time spare. Yet change always happens, and new work always emerges unexpectedly. So either they have to ignore the new work and press on with what they were doing, even if the new work would have been more valuable, or more likely they have to put down the thing they were working on to work on the new more valuable work, creating delay and waste in what was previously being worked on. Wider plans start to fall apart as the people who needed something doing in order to get their thing done now have to wait themselves, and so the impacts of the change ripple out across the system.

For an analogy, think of a road that's 100% utilised. Every piece of the tarmac is covered by a car. If each car carries on moving at the same high speed, everything is fine. But the minute one car slows down, or changes lanes, the cars behind it have to slow down too, which means the cars behind them have to slow down, which is an effect that then ripples down the lines of cars, until eventually the whole system grinds to a halt.

This is how traffic jams actually happen. They're not caused by one car stopping in the road, they're caused by roads with too many cars on experiencing some, often small, variation in the speed of the flow, which has knock on effects down the line until everything crunches to a halt. So it's surely far better to leave some space between the cars, and run the highway at less than 100% utilisation, so that if one car does slow down, it has time to speed up again before it gets in the way of the cars behind it.

The same thing happens in organisations when people are 100% utilised. In a system running at 100% utilisation, a small variation in one area often ripples through the rest of the organisation, leading to gridlock.

100% resource utilisation also causes harm in other ways. If people are 100% utilised, they have no time for creativity and innovation. After all, in order for you to know that someone is 100% utilised, you need to plan up front everything they're going to be doing. Just like your school did when it handed you your timetable at the start of the year. Only the difference was that your school knew all of the things it had to teach you that year to get you through your exams. If you're working in lean and agile, you're saying that you can't know all of the things you need to do up front, so you need to be able to change and innovate as you go along. 100% utilisation assumes that all the work can be known up front, removing any ability to change and innovate.

From a human perspective, 100% resource utilisation causes burnout too. Think of your own computer. If you open lots of different programs on your computer at the same time, and have two Internet browsers running at the same time, each with dozens of tabs open, does your computer run smoothly and speedily, or does it get hot, with its fans whirring loudly and slow right down? People are no different. Work them at full tilt 100% of the time and eventually they slow down and burn out. You may get short-term productivity gains, but if you look at the amount delivered in total over the medium or long term, it's likely to be significantly less. It's the reason one of the principles of the agile manifesto talks about people maintaining a sustainable pace. The ideas of working at sustainable pace found in the agile manifesto very likely comes from the Extreme Programming framework, and in his seminal book on the topic, Kent Beck has two lovely quotes that sum up what sustainable pace really means.

> *"There are other human needs, such as rest, exercise, and socialization that don't need to be met in the work environment. Time away from the team gives each individual more energy and perspective to bring back to the team. Limiting work hours allows time for these other human needs and enhances each person's contribution while he is part of the team."* (Beck, 2005, p24)

> *"Software development is a game of insight, and insight comes to the prepared, rested, relaxed mind."* (Beck, 2005, p41)

Just like a computer being maxed out, working at 100% utilisation removes people's abilities to rest, relax and recharge their minds, leading to lower quality work, delay and burnout. Sometimes you even get a perfect storm, where 100% utilisation leads to someone burning out, which slows their work down whilst they take time off to recover, which has the 'traffic jam' knock on effect across the wider system, slowing everything down. An entirely predictable and avoidable situation, but one that occurs all too often.

So what's a people manager's role in all of this? Well, very often people managers are involved in the sorts of resource management issues we've just been looking at. Even if they're not directing teams and 'resources' themselves, they often help set the policies and procedures for how an organisation uses its 'resources', or 'people' as they're called outside of the workplace. In a world where we're brought up to believe in 100% utilisation, shifting this approach will be extremely difficult, but it really could deal with so many problems in delivery and innovation that an organisation is experiencing. The importance of people managers in helping this shift to happen is likely to be immense.

Idea 9. How To Move Your Lean-Agile People

> **Hypothesis**: Moving or reallocating lean-agile people as if they were 'resources' is likely to kill their motivation or lead to their resignation.
>
> **Experiment**: Review your current approach to moving people within your organisation, to see if it truly respects their sense of autonomy and desire to collaborate widely.

A while ago I was chatting to a good friend, someone who is easily one of the best lean-agile practitioners that I know. We were catching up on work, and she told me an interesting story that got me thinking about another angle of lean-agile people management.

As I say, this friend is fantastic, so it's perhaps no surprise that she'd become well known across her organisation, and her talents were much in demand. Being an agile person, she was always happy to talk to people too, running the odd internal workshop or informal coaching session across her wider organisation for the benefit of the organizational system as a whole.

One workshop she ran seemed to catch the eye of senior people in another part of the organisation, as they invited her to come and work for them instead. Committed to her current teams, she thanked them and said she'd always be happy to chat through any issues they had, but that she had to stay where she was and see through what she'd started, whilst keeping her personal work-in-process limits nicely respected.

Then shortly after that, the overall director of her entire department forwarded her an email, asking what was going on. It seems the senior people that had tried to poach her had approach her bosses' bosses' boss with a formal request that some of her time be allocated to the new department to help them.

Can you imagine her reaction?

Here was a person who's hugely talented at implementing lean-agile approaches, suddenly being treated like a simple 'resource', merely a productive unit of value that can be moved around at the request of hierarchical authority, without even being informed or consulted, and despite the fact that they'd already said no.

First, it severely damaged the sense of psychological safety she had in the workplace. Having been happily doing what she believed to be a good job, she was suddenly thrust into an uncertain situation over which she seemed to have no control. With her stress levels raised, she had to work even harder just to maintain the same performance.

Second, it damaged her intrinsic motivation. If you read Dan Pink's book 'Drive', he convincingly makes the case that knowledge workers, people who know more about their work than their bosses, are typically motivated by internal factors such as having autonomy over their work, and the ability to master skills in their chosen field, rather than by extrinsic motivators such as hierarchical command. She was certainly a knowledge worker, someone who knew more about the work she did than her boss, but here was someone trying to 'extrinsically motivate' her by escalating to her higher authorities the fact she declined to move.

Third, it made her even more determined not to move. For if that was the way this new department treats their employees and tries to enforce their compliance, they wouldn't have been a good fit for the type of change role she was looking to do at that time. In fact, it gave her a pretty clear sign that this department was not yet ready for lean-agile adoption, if that was the way its senior leaders thought they could treat people.

So as a senior leader, or a people manager tasked by a senior leader to move a lean-agile 'resource', what do you do in this situation? Genuine lean-agile talent is in very short supply, and demand is currently high.

In a way, you can't blame that department for doing what it has always done when faced with an opportunity. Invoking hierarchy, escalating to senior management and optimising for its part of the system. As a non-agile place, it doesn't yet know to do any different.

This is where the role of agile people managers could really come into its own. Lean agile people are generally open, collaborative types who are happy to help wherever they can be of use. Traditional behaviours taking place around them can kill that stone dead, but with people managers who understand both sides of the picture, the situation could be so much better. You've got people who want to help, people who want to be helped, all you need is some people in the middle who could help facilitate the relationship between the two different worlds.

If you're reading this, and you work in people management, this could well be you.

Section 4: Hiring

Idea 10. First Impressions Count When Hiring Lean-Agile Talent

> **Hypothesis**: Genuine experienced lean and agile talent is very in short supply.
> **Experiment**: Relentlessly test and optimise small changes to your hiring process and measure the impact on talent attraction and hiring.

I spend a lot of time on LinkedIn. I see a lot of 'lean / agile' job advertisements, and I receive a lot of speculative job applications from fellow lean and agile practitioners. There is one common thread I see present in all of them.

For all that there are many thousands of people out there claiming to be 'enterprise agile/agility coaches' or 'lean/agile coaches', often with the word 'senior' mixed in there too, there are in reality very few people who understand lean and agile to a meaningful degree. Interestingly, these people who do understand lean and agile tend not to be the ones that use the words 'enterprise', 'agility' or 'senior' in their personal titles, making them harder to spot. Being talented, they also tend to be in high demand, making them hard to attract in the first place. In essence, my personal experience is that if you want to attract truly experienced and capable lean and agile talent, you have a real job on your hands.

However, as the old saying goes, you only get one chance to make a first impression. It's true of people, but it's also true of organisations. When hiring lean and agile talent for your organisation, have you ever stopped to think that your hiring process might be the first impression your future employees get about your organisation, and how that might be setting the tone for their entire time with you?

There are lots of different ways your hiring process could be giving the impression that your organisation isn't a happy place for lean and agile practitioners, or isn't somewhere lean and agile are expected to thrive. Let's look at a few of them, and how you could stop them happening.

The first thing of course is the advertisement for the job itself. This is pretty easy to get right, just get someone internal who understands lean and agile to write it, but it's surprising how often organisations get it wrong. As some quick checks, make sure you're spelling agile terms correctly (it's Scrum, not SCRUM, because it's not an acronym), that you aren't calling agile a process or methodology, and aren't asking for irrelevant certifications like PRINCE2. I ran some research into this in early 2016 via my www.agileforrecruiters.com project, and practitioners said very clearly that badly written job ads were often their number one way of spotting which organisations not to apply to.

Assuming you get the job ad right though, what if the first round of your hiring process is some online screening questions? Many large organisations use these, and they seem like a tempting way to sift and filter large volumes of applications. However, one of the four key elements of the agile manifesto says that we value;

"Individuals and interactions over processes and tools." [5]

Filter a candidate in or out on the basis of an automated test before they've even spoken to anyone, and you're strongly suggesting your organisation doesn't get this part of the manifesto. Besides, anyone with half a mind can game these tests pretty easily. They don't add value, and they shout that you don't get agile. Personally, I wouldn't use them.

The same goes for CV scoring software. Not everyone knows about this at the moment, but there are programs out there that scan CVs for keywords and score them on the candidate's suitability for the job. I

---

[5] http://www.agilemanifesto.org

can't think of anything much less agile than these systems, or anything much more ridiculous. They just turn job applications into tests of a candidate's ability to get the right keyword densities in their CVs. The candidate may even have outsourced this task to one of the many companies that you can pay to get your keyword densities right. These systems really are a terrible idea in lean-agile hiring.

Say though that you don't have those tests, that you don't use CV scoring software, and that all applications are read and screened by real people, who then draw up a shortlist of candidates to interview face to face. Great stuff. However, what if that interview is then completely formulaic? One where the interviewer is given a fixed set of questions they have to ask, drawn from a pre-approved question bank, and candidates have to 'give an example of a time when they have…', in order to be scored out of five on each answer. Again, you're valuing processes and tools over individuals and interactions. Not to say each interview should just be a freestyle random chat about just anything, but it's not difficult to create a fair and equal interview for each candidate that's based around human conversation rather than reading out a written exam.

Say a strong candidate gets through all of this though. You don't filter with screening questions, you let real people draw up shortlists, and you run human interviews face-to-face. Brilliant, but there's still one big thing that could ruin the fantastic first impression you've been giving; your reference checking process.

Principle five of the agile manifesto says;

> *"Build projects around motivated individuals. Give them the environment and support they need, and trust them to get the job done."* [6]

Yet nothing says 'I don't trust you' like a reference checking process.

Of course, there's a problem here. On the one hand, taking references

---

[6] http://www.agilemanifesto.org/principles.html

is all about a lack of trust, otherwise you'd just take at face value what a candidate has written on their CV and said at interview. On the other hand, it's a process that probably does still need to exist. If no-one ever checked any references, I'd be applying to jobs as the former CEO of Apple by now, at least. So how do you square this circle?

Well one simple way to do it is to make the referencing process as simple as it can be. If you insist on checking every single detail a candidate has put on their CV, wanting certificates from every online training course they've done, or demanding a reference from every unpaid voluntary role they've ever done, no matter how short, then you're screaming that your organisation isn't built on trust. Perhaps just check the references that matter, and only ask to see the more significant certificates, like their degree.

A truly agile organisation could go further than this though. Agile people love feedback. We also use a style of leadership called servant leadership, and one of the ten principles of that is empathy. So how about you turn the reference checking into an explicit session of feedback and empathy?

Rather than contacting previous employers just to check the facts of what a candidate has said, use it as an opportunity to gather some feedback for the candidate too. You'd need to agree it with the candidate first of course, but if you did, you'd be demonstrating clearly that you hold true to agile values such as transparency and collaboration, working with the candidate right from the start to help their personal growth.

On top of this, another element of empathy is understanding the experiences that have shaped the people you're working with. So perhaps you could use the reference checking process to get a sense of the organisation or organisations the candidate has been working for previously. So you wouldn't just be finding out about the candidate, you'd be finding out about their previous employers too. Getting a sense of the experiences and environments that have shaped the candidate will really help build empathy with them once they join your organisation.

Additionally, doing both of these things gets round another big problem that reference checking can cause. Sometimes you come out of a final round interview all pumped up after a great conversation about an exciting role with some people you love to work with. They offer you the role, and then you have to stop talking to the people whilst the references are checked and you find out if you will actually be working with each other. Using the reference process to build empathy and help the candidate grow gives you two excellent reasons to keep those conversations going, meaning that once the candidate does start, the conversation and collaboration is already open and strong.

Great agile talent doesn't grow on trees, and in many sectors it's a seller's market right now, with the great candidates holding most of the cards. Organisations that work hard to give the right first impressions to candidates, and demonstrate that theirs are workplaces where lean-agile values are lived and breathed every day, are far more likely to attract the best talent. Your hiring process is a huge part of your first impression. This is not to say that changing hiring processes is easy. There's often a lot of organisational policy and relevant employment law tied up in there too. However, wherever there are large problems, the answer is often to recognise them, then make small, incremental but continuous improvements. So have a think about what your hiring process is really saying about your organisation, gain genuine feedback from both successful and unsuccessful candidates, and then run small experiments to improve it continuously. Even a small change might make a big difference. Hopefully this idea has given you some areas to get started on.

Idea 11. Hire Lean-Agile Leaders

> **Hypothesis**: Training existing leaders in lean and agile approaches may be necessary, but it may not be sufficient.
>
> **Experiment**: Scope out the need and feasibility of making a senior level appointment with specific responsibility for lean and agile within the organisation.

For your lean-agile transformation to succeed, you need to have leaders who understand what it is and what it is they must do. As W. Edwards Deming said;

> *"Support of top management is not sufficient. It is not enough that top management commit themselves for life to quality and productivity. They must know what it is that they are committed to — that is, what they must do. These obligations can not be delegated."* (Deming, W.E., 2000, p21)

However, if you think about it, how serious is your organisation about having senior leaders who know what it is they must do when it comes to lean and agile?

You see, over the years, a lot of people have started becoming 'agile leadership coaches'. The idea being that these people can take your existing leaders and turn them into people who live and breathe agile values and principles, and fully understand which practices are and are not helpful in bringing those values to life. It's a situation where everyone wins. The leadership coaches keep themselves in jobs, and the leaders get to keep their jobs too, despite their organisation changing around them.

Now I'm not saying this is a bad thing. Time and time again I've seen the main gap between leadership and agile teams simply be one of communication. Both of them want to see quality improved, delays reduced and workplace happiness increase. Sometimes all it takes is explaining to the leadership that these are things that lean and agile can help to make happen, and how they do so. Training and coaching leaders in lean and agile is hugely important and necessary.

However, whilst leadership coaching and training is necessary, is it sufficient?

A few years ago I was working with some senior people in the UK marketing industry, helping them to understand how agile might apply in their context. Whilst we were chatting, one of them got to talking about how much the marketing industry had changed once marketers were finally allowed a seat on the company board. Up until that point you had CEOs, CFOs and various other types of Chief Officer, but there typically hadn't been a board level 'Chief Marketing Officer' position or its equivalent, whatever name it may have been given.

At the time that struck me as strange, given how essential marketing is to the success or failure of a company. So, what if lean and agile turn out to be as important to some companies today as marketing was discovered to be a few decades ago? As I've said, training leaders is great, but when companies decided that marketing was sufficiently important to warrant a seat on the board, did they train up existing CxO leaders to be their chief marketing board member, or did they find someone who had lived and breathed marketing for decades already? In almost every case I'm certain the option chosen was the latter. If it was, then why should lean and agile be any different?

Now, my guess is that very few organisations, if any, have given lean and agile a seat at the top table. The 'leadership coaching' industry seems to be filling the gap for the time being, trying to turn existing leaders into agile leaders. However, I suspect a greater reality is that in many instances, the leadership coaching industry just isn't sufficient to reach the tipping point for a meaningful and sustainable lean-agile transformation. Leaders are busy people. Ask any agile practitioner you like, and I'm sure they'll tell you a story of the time they tried to run two days of leadership training, and were asked to cut it down to two hours. Ask those that did agree to cut it down to two hours just how many times those leaders then cancelled their attendance at the last minute.

The reality often is that agile is seen as primarily a team thing, that leaders don't really need to concern themselves with it too much, and at best can just be briefly trained in to explain what the teams are now doing, before they go back to their everyday jobs. The problems then are huge. The teams see that there's no real organisational buy in or sense of urgency for the change, and the leaders, not knowing what it is that they don't know, can push old practices back into the teams, slowly eroding their agility, causing all the agile talent you hired to resign, and bringing your whole transformation to a halt.

So if in many instances leadership coaching is necessary, but not necessarily sufficient, then what should an organisation do? For me, the answer seems simple. Add in a lean-agile leader role to the upper levels of your management structure, and hire into it someone with substantial experience in lean and agile. This is not to say agile should therefore become some separate department or branch of your organisation, staffed solely by agile people who only talk to each other. Agile is something that has implications for every part of your organisation, and its something every leader needs to understand and buy into. This is the same scenario as there being a senior marketing person in the organisation, but anyone in a customer facing role needing to have an understanding of the organisation's brand identity.

However, amongst these agile generalists, might it not be worth having an agile specialist leader? In marketing terms, the person who would oversee the creation of your brand identity and set the direction in the first place? Who else is going to have the time to monitor the progress of your transformation, and be held accountable for its success? If you're going to have Scrum Masters, who are they going to report into to ensure that Scrum (or whatever framework you choose) is actually being followed, and isn't be compromised in a thousand different directions?

You can teach agile to leaders as much as you want. You can coach them over time to build their understanding. But unless you get lucky and find an existing leader who decides to buy into it so strongly that they're prepared to give everything they've got to getting it right, you might well have a significant gap at your top table. Perhaps it's time you added lean-agile leadership into your senior level resourcing and hiring strategies.

Idea 12. Stop Hiring Great People From Great Companies

> **Hypothesis**: Experience within and understanding of a sector is as important as experience and talent in a specialism.
>
> **Experiment**: When hiring, closely evaluate your balance of factors between candidates' experiences of working in new environments and their experience of dealing with legacy issues in your industry sector.

What a ridiculous title or this idea right? Of course you want to hire great people, why wouldn't you, and if they come from great companies, then surely they'll know the most about how to make a company great?

Well, yes and no…

All too often I've seen organisations that want to move to more lean and agile ways of working splash out lots of cash on hiring people from the famous digital organisations. Naming no names, but they're typically the organisations that people most commonly associate with life on the Internet these days. Now, these people are often great, and they've seen inside the workings of some of the most wealthy and successful companies on the planet right now. But that doesn't mean they know how to make *your* organisation great in the digital or agile space.

The thing people overlook when organisations hire great people from great companies is that companies are often very different from one another. For example, most of the modern tech giants were only founded in the last 15-20 years, and some of them are even younger than that. They've been hugely successful for sure, but fundamentally they had a blank canvas to work with. No legacy mainframe database systems, no legacy customer base, no legacy organisational culture.

They could start truly from scratch and focus on what's needed to be successful in the modern age. They never had to undergo a lean and agile transformation, as they could just start that way from scratch.

Perhaps the most famous example of this is Spotify, who created the 'Spotify model' for how they do their work. It's become quite famous as a model amongst certain parts of the lean and agile community, but the clue is in the name. It's the Spotify model, because it was created at Spotify and it works at Spotify. Take it and apply it to a different organisation, especially one that faces many issues Spotify never had to deal with, having started from scratch, and I'm certain you'll not get the same results that Spotify got.

My guess is that your organisation isn't Spotify. If it's been around a decent length of time, it probably has legacy systems to deal with, a legacy code base to clean up, and almost certainly a legacy organisational culture. What you probably need is an approach that's proven to deal with these issues, not an approach that's never had to deal with them. So if organisational models don't necessarily transfer well between organisations, why would organisational leaders be any different? Do these great people from great modern digital companies have any experience of dealing with legacy systems, code bases and cultures? Quite possibly not, and yet, these areas are often the issues a lean-agile transformation needs to tackle the most.

Let's take an example from a completely different area. When the band Metallica was looking for a new bassist back in 2002, did it go out looking amongst folk and country music fairs across the nation? Of course not, but why not? There are certainly loads of great bassists out there amongst the folk and country music worlds. Fundamentally they all play the same instrument too, the bass guitar. Metallica didn't look the worlds of folk and country music because they needed a bassist who could play heavy and thrash metal, and who had experience of 'working in that sector' as it were. In the end they hired Robert Trujillo, who used to be the bassist for Ozzy Osbourne of Black Sabbath fame. Someone who had experience of Metallica's style of music and culture.

The same principle likely holds true when people are hiring for their organisations. There are many great people out there amongst the great companies of our modern age. They may be highly skilled at setting up new companies and building them from small startups to world dominating enterprises. Even if they weren't in the company right from the start, they may still be great at achieving results within a relatively new enterprise. But are they the people who can do the type of organisational change your organisation will need to do? Do they have experience changing and improving legacy systems and cultures, or are you at risk of hiring a bassist that plays a completely different style of music, which will leave neither side happy or successful?

Idea 13. What To Look For In A Great Lean-Agile Hire

> **Hypothesis**: Many standard hiring approaches are designed to look for non-agile skills and behaviours, and unwittingly prevent such behaviours from being displayed.
>
> **Experiment**: Evaluate how your current hiring process draws out a candidate's authenticity, transparency, honesty, humility, inquisitiveness and humour.

What do you look for in someone you hire? Is it them meeting a predetermined set of criteria? Is it their performance against various trait and personality tests? Or is it some intangible quality that just makes you think they're the right person for the job?

I'll be honest, I'm not a big fan of the idea that people have inherent traits. It's largely a product of 1930's management thinking, and it's an approach that believes people have inherent traits within them, either learned over time or present from birth, that determine how they will behave in certain circumstances. On one level it makes intuitive sense. People do seem to have typical ways they behave, typical things they do when certain situations occur, and typical patterns of behaviour they fall into.

However, to go too far down this line of thinking is to put at risk a lot of the human side of agile. People don't exist in a vacuum, they're shaped by and they shape their environment on a daily basis. To see people as defined by their traits is to risk missing the influence of the system around them. As W. Edwards Deming said:

"A bad system will beat a good person every time."
(Deming, W.E., 1993)

Too often, believing in trait based approaches to defining people allows us to imagine that failure is due to the failings of individuals, and ignore the failures of the system in which those people are working.

However, if we ignore the individual completely, then we end up pretty stuck when it comes to hiring. If success is all about the system, then surely it doesn't matter who you hire or what they're like, does it? Well of course it matters. We just shouldn't believe that how someone behaves is fixed, unchangeable and not able to be improved through a focus on empathy, compassion and learning.

With that in mind, here's what I've come to notice over years of hiring people for agile roles, in terms of the behaviours great agile people seem to learn and exhibit.

The first is a love of failure. Some people have been taught by bad systems to manage the message at all times, to only ever talk about good news and to avoid or cover up failure at all times. Great agile people tend to be the opposite. They understand that if you aim to avoid failure, then you're aiming to avoid innovation and learning, and you never make any progress.

I remember once a person I was working with made a potentially huge error. They caught it quickly and the potential downsides didn't materialise, but if they had, the impact would have been huge. Rather than cover it up though, they came to me white as a sheet and shaking as soon as they realised what they had done, openly admitting what had happened and suggesting ideas for how to fix the problem; a fantastic response. Once we'd dealt with the situation, I went out and bought them a bottle of champagne on the spot. After all, failure's great when it creates valuable learning. I knew this person was never going to fail like this again after what had happened, and that was a cause for celebration.

Another great behaviour is authenticity. It's hard to put authenticity into words, as it's all about being who you truly are at all times, and who you are can vary hugely between people. You notice authenticity in lots of different ways though. In their language, authentic people tend to speak like human beings rather than corporations. They say that they're happy rather than 'comfortable', they say that they're annoyed rather than 'disappointed'. They bring their whole selves to work, with all the benefits that brings in terms of creativity, diversity, enthusiasm and communication.

Following on from this is honesty and openness. There's a joke doing the rounds online like this that I love. There's a person sitting in a job interview, and the interview panel asks the person what their greatest weakness is. The person says *"Honesty"*, to which someone on the panel replies *"I don't think honesty is a weakness"*. The person just shoots back with *"I don't give a **** what you think"*.

Of course, that's an extreme example for the purposes of a joke, but if I were hiring someone into an agile team, I'd want to know that they were unafraid of sharing their opinions, and were capable of being honest and open with those around them at all times.

That said, humility is another important behaviour that allows honesty and openness to work well. A humble person knows they're not an expert, which is essential for agile practices such as inspecting and adapting and responding to change. A lack of humility can breed an over-confidence that dominates the team and prevents it learning and optimising its ways of working. Agile needs servant leaders, and it's hard to serve others without possessing humility yourself.

A sense of inquisitiveness arises from this humility too. If you know you're not an expert and never can be, then you're going to be far more interested in constantly learning, changing and adapting. So what if someone got an MBA five years ago? What did they learn last year, last month, last week? Great agile hires may have a lot of certifications and letters after their name, but they often only see these things as the happy by-product of what matters, a constant quest for new knowledge.

With this inquisitiveness often comes excitement and positivity. Great agile hires often talk about their work as something exciting they have done that they want to share with other people. That said, they very much don't display the unthinking positivity that's poisoning so many large organisations at the moment. If something was rubbish, they'll say so. If they didn't, they wouldn't be being authentic and honest. But the fact that it was rubbish, and the way it presented an opportunity for learning and improvement, is something they'll find really exciting.

All of these behaviours together often bring one final behaviour that can really help make agile work: a sense of humour. If you're authentic and open, you're not afraid to laugh. If you're humble, you're not afraid to laugh at yourself. If you're excited about your work, then that excitement often spills over into laughter. Equally, sometimes a sense of humour is an essential requirement for people working on agile transformation, as it's the only way to stay sane in the face of repeated challenges.

In and of themselves, all of these things sound nice, and not too controversial. If you think about them in practice though, often they run counter to how we normally hire. In an interview, people want to give the best possible impression of themselves that they can. So ideas like admitting failure and being authentic go straight out of the window, along with honesty and openness. Humility often follows quickly behind them, as the interviewee feels like they have to take the credit for things that have gone well around them in their career to date. Inquisitiveness is difficult at interview too. A genuinely inquisitive person may be far more interested in learning from the interviewer's opinions than in just sharing their own. You might like someone's excitement and positivity whilst interviewing, but if that spilled over too often into humour, would you not just think that they weren't taking the role seriously?

Great hiring in agile is integral to the success or failure of an agile adoption. Yet the ways we see the conventional business world, and the ways we think we need to behave at interview often run counter to so many of the behaviours exhibited by great agile people. If you wonder why you keep on getting the same results, maybe you should look at the behaviours your hiring process actually encourages.

Idea 14. When Hiring Agile Talent, Look For People Who Have Done A Bad Job

> **Hypothesis**: Great lean and agile people may have chequered career histories, littered with employment gaps and failures.
>
> **Experiment**: Relax some of your hiring conventions that might automatically discount great change agents due to employment gaps, short-term engagements or poor references, and focus instead on discovering what the candidate learned from these experiences.

What's the one thing you look for when looking to hire someone into an organisation? Great qualifications, evidence of previous successes, fantastic references and testimonials on LinkedIn? Now what do those things all have in common? They point to someone who has been good not bad, who has won friends rather than alienated people, who has been successful and hasn't failed. Or at least, if they have failed, they've done well enough to cover it up and move on.

Perhaps though, we should step back and think for a minute about what doing a good job means. A good job is always defined in the context of the organisation in which the job is happening. So if someone has done a good job in an organisation that is similar to yours, then it seems reasonable to suppose that they will do a good job in your organisation too.

However, if an organisation is moving to a different approach such as lean and agile, this will mean that the organisational context will be changing too. As a result, some of the activities that were previously classed as doing a good job will now classed as doing a bad job, and vice versa. For example, it may have been that people were promoted to more senior management positions when they could demonstrate that they were good at directing and controlling people, that they could define and implement the process people should follow, that they could 'drive teams', that they could centralise decision making with themselves and take the tough calls that others would not. However, with a shift to lean and agile, many of these previously rewarded behaviours should now be being discouraged. Instead, we want to decentralise decision making down to teams and individuals, we want people to be intrinsically rather than extrinsically motivated, and we want to empower people to work in the ways they think are best.

Suddenly, the people doing a good job are doing a bad job, and the people who previously were seen as doing a bad job, that refused to control and direct others, may find that their time has come to shine. Given how recently the lean and agile movements have become fully mainstream, it's likely that the talent you're looking for, if they've been truly living the values for a long time, may well have pretty chequered CVs and work histories, without the continual successes you'd normally be looking for.

To think of an analogy for this for a minute, I don't know if you studied the story of the peppered moth when you were at school, but in case you didn't, it went like this. The peppered moth comes in two forms, a largely white-bodied moth, and a largely black bodied one. Before the industrial revolution, the white-bodied moth was more common, as it was better camouflaged against leaves and trees. However, when the industrial revolution happened in the UK, suddenly the black-bodied form of the peppered moth became more common, as the trees and leaves became stained black with soot from industry. Once the intense soot pollution of industry started to be reduced, the white-bodied moth became more common once more.

The same is true in many respects for agile people. The people who prefer to work collectively in teams and to support and empower others may not have done as well in large organisations set up on a command and control basis, but now organisations are moving away from command and control to more lean and agile approaches, they are beginning to thrive.

On top of this, there's the issue of the people who are out there on the forefront of organisational change, genuinely trying to move people, teams, departments and organisations away from their old ways of working to more lean and agile approaches. This is a tough job. It is a really tough job. Creating genuine, deep rooted and lasting organisational change is immensely difficult and hugely exhausting. Some people I've met and interviewed over the years doubt it can ever genuinely be done, and they do have a few strong arguments on their side. Still, many people, myself included, enjoy trying to bring about true organisational change, and give it our all.

Given how difficult and exhausting it can be though, how mentally fatigued and demoralised you can become through trying to shift the inertia of an organisational status quo day in and day out, sometimes these people fail. Sometimes the organisations just aren't ready for the change, or the odds are stacked too far against the change agent. Sometimes, truth be told, some of these change agents fail too, and burn themselves out or need some time off to recover. To step out and lead the way to a new way of working takes courage, determination and fantastic interpersonal skills. Just the sorts of things you should be looking for in great hires, but as a result of the pressure and stress, it may well be that these people have somewhat chequered CVs and reference histories. It may be that an organisation they were trying to change had forced them out, because as Peter Senge observes in his work on systems thinking;

"The harder you push against a system, the harder the system pushes back"
(Senge, P.M., 2006, p58)

So whilst it's possible that some people who have been classed as doing a bad job were just bad at their jobs, I'd argue that it's far more likely that they were doing the best job they possibly could, possibly a brilliant job, they were just doing it within a system that didn't think that what they did was good.

So when hiring agile talent, it's perfectly possible that the best people you can find will be the ones who look like they've been doing a bad job, or have gaps in their employment history, or have jobs that give them poor references. If your organisation is making a change to lean and agile ways of working, then quite possibly it's these people that will make all the difference to your success. The trick is not to discount them straight away, because they don't fit the profile of what successful hires used to look like, but instead to see their 'failures' as potential examples of doing a great job.

Bring them in, talk to them, ask them what they learned from the experience and what they'd do differently next time. If they've no self-awareness or don't seem to have learned a lot, then perhaps they are just a bad hire. But a great lean and agile person will likely not be able to stop talking when asked about these things. The people you want are the ones who see their failures as the opportunities for fantastic growth and development that they are, and are clearly hungry to get back on the horse that threw them off, and ride it even better next time round.

Idea 15. Create Genuine Lean-Agile Job Descriptions

> **Hypothesis**: A move to lean and agile ways of working will require job descriptions to be rewritten, and there are few if any good options currently available to buy in from the open market.
>
> **Experiment**: Proactively run an exercise writing job descriptions that suit both lean and agile, and your organisation's industry context, bringing together experts from the worlds of lean and agile, along with other relevant SMEs, in order to do so.

If your organisation is going through any sort of lean-agile transformation, you'll have come across this issue I'm sure. It's a pretty simple one on the surface, but one that's hugely difficult to resolve in reality.

It stems from the fact that different agile frameworks create different roles within teams. Roles such as Scrum Master and Product Owner in the Scrum framework, or roles like Release Train Engineer and Epic Owner if you're looking at the Scaled Agile Framework. At the simplest level, you need to create new job categories to enable these roles to exist. Simple as that is to say, there's potentially a huge amount of complexity in it.

Before we step into the difficult parts of this though, we should first ask whether this whole thing really needs to be difficult. Can't you just keep your existing job categories and role descriptions, things like Project Manager and Programme Manager, and let the teams call the people doing those roles whatever weird new name they want? After all, surely the same sort of work will still need doing, even if the name of the role changes?

Well, no. Whilst this approach may seem simple and intuitive in principle, and I know many organisations have tried to take this route, it's an approach that only ever stores up pain for the future.

You see, formal job titles and role descriptions matter and have impact. If they didn't, you wouldn't have them in the first place. So if someone for example becomes a Scrum Master, but has the existing Project Manager job category and description applied to them, then sooner or later there's going to be a conflict they have to resolve. For as much as they may want to empower the team for which they are the Scrum Master to take decisions and self-organise, their Project Manager job description will be expecting them to be 'driving' the team and being accountable for taking decisions on their behalf.

So which master do they serve? The rules of the job they're meant to be doing, or the rules of the job that their employment contract states they should be doing? Ultimately a situation of changed informal role titles and expectations without changes to formal job titles and descriptions becomes untenable, and can also cause wider harm to your lean-agile transformation. After all, not only will someone end up conflicted as to what they're being expected to do, the situation also clearly sends the message that nothing's really changing in the transformation. That it's just about giving people and activities new names, without changing the reality of what they are.

So at some point, you're going to have to bite the bullet and create new job categories and job descriptions. However, this is another place your lean-agile transformation can come unstuck, as to undertake this task, you're now looking for a pretty unique sort of person. One who understands lean and agile, and also understands how to write role descriptions. For if you get it wrong, you create huge problems for your transformation. What may previously have been misunderstandings and anti-patterns about what agile roles do on a day-to-day basis now become the official, sanctioned truth that people must follow, because their job description says so. Very often, organisations attempt to write new role descriptions, but the people doing so don't fully understand lean and agile, so the description ends up as a mishmash of lean and agile words in amongst traditional project management responsibilities.

You end up with a compromise that no one is happy with, neither from the old world or the new.

On top of the complexity of rewriting role descriptions, and the need for precision, you often also get the problem that organisations find it hard to change their role descriptions for all manner of reasons anyway. Do you need senior level approvals to get anything changed in 'the system'? Do you need to consult with staff and / or trade unions before making any changes? From a human perspective, how much additional uncertainty and fear do you risk creating by chopping and changing what people's roles mean?

It's perhaps for this reason that some organisations just decide to buy in their role descriptions from elsewhere. After all, as the saying used to go 'No-one ever got fired for buying IBM'. In the sense that if you buy in something everyone else is using, then surely you can't go wrong, or at least you can't get the blame when things do go wrong. The problem here though, and I've experienced this first hand more than once, is that often these externally provided role description frameworks haven't kept pace with the realities of lean and agile ways of working. So either they present a confused mess of information, or they flat out reject many of the core elements of lean and agile working in what they state various roles need to do. For example, how do you make continuous testing of what you're producing integral to your delivery process when your bank of role descriptions contains references to a distinct pre-launch testing phase, and roles that are solely responsible for running that phase?

For one final piece of complexity, we also have to think about how role descriptions might also feed into other functions within an organisation. For example, if an organisation works in a regulated industry, then it is likely that there will be an organisational framework for ensuring people comply with those regulations. As part of this, certain types of role will be held accountable for ensuring regulations are met. Typically, this approach really struggles with lean-agile ideas such as decentralised decision-making, team autonomy and collective accountability. So this whole role description issue isn't just one that affects HR departments. It's one where HR departments are going to have to collaborate with any new regulatory models that are emerging within the organisation, and ensure that they stay true to lean and agile values in their role based implementation. That's a lot of plates to keep spinning at any one time then.

So how do you fix this? Well, the first thing to do is not to try to shy away from it or fudge the whole issue. It's going to come back and bite you at some point, so you may as well get started on it early. Papering over the cracks with changed job titles without changing role descriptions is only going to cost your more pain and rework in the longer term.

Second, make sure you engage people who know what they're talking about in their space. The role descriptions you write are going to have huge impacts across the organisation, from how people behave, to whether people want to work at the organisation at all given the way roles will be described on advertisements during the hiring process. Get things wrong at this stage, and you could undermine the entire transformation.

Third, collaborate. I've picked out regulatory compliance as one area that might be affected by role descriptions coming from HR, but doubtless many others exist. The role description is at the heart of how work will be done in an organisation, and if it can meet the needs of the organisation, whilst also staying true to lean and agile values and principles, then it will be absolutely invaluable to the success of the organisation's lean and agile transformation.

Idea 16. Experiment With User Story Job Descriptions

> **Hypothesis**: Traditional job advertisements are effectively 'big up front design' as found in traditional project management approaches.
>
> **Experiment**: Try writing job advertisements as collections of user stories instead, giving candidates the space to say how they'd answer them, and even letting candidates re-prioritise them against each other too, based on their prior experiences.

I'm really interested in how people get hired for jobs in lean, agile and their associated frameworks. Whilst lean and agile have transformed so many workplaces and ways of working, they still don't seem to have touched many recruiters. Time and again you see adverts for jobs in lean and agile that look just like traditional jobs ads, with an overview of the role, a list of tasks the role is required to perform, and a list of experiences, skills and/or attributes the successful applicant ought to possess.

This works fine in some environments, but I'm increasingly convinced it doesn't make any sense for recruiting for roles in lean and agile, or indeed any sort of innovative workforce working in an environment of complexity, change and uncertainty.

First of all, a lot of this approach implicitly involves up front planning and design. With a traditional mindset, perhaps even a small committee of people, you get together and write down all the things you think the person will need to do, and specify them in a big list. Exactly the same way a traditionally delivered project would specify all of its requirements before moving into its construction phase. Yet we know this project approach fails in complex, fast-paced and uncertain environments, that's why lean and agile are now so popular as alternatives. So if you're hiring for a role in a complex, fast-paced and uncertain environment, isn't taking the 'big up front design' approach of a traditional project equally as problematic? How can you possibly know what the person will and will not need to do, especially before they've even started, the time when you know the least about what will happen further down the line?

Another problem with this is what I call the 'eight legged horse' problem. In Norse mythology, the god called Odin has a horse called Sleipnir, which has the remarkable benefit of having eight legs. Obviously that makes it a pretty good horse. The problem is of course, in the non-mythological world of us mere mortals, no brilliant eight-legged horse has ever existed or will ever exist. The reason I mention this is that very often, if you try to specify up front what you would like a role to do and what sort of person is needed to do it, then your lack of knowledge about what the reality might actually be means you over-specify the role description, and create a role and person specification that describes a role that could never be done and a person who could never exist. You advertise for your own version of an eight-legged horse.

This is of course assuming that you even do all of this thinking. Quite often you see people's lack of knowledge around lean and agile result in them simply copying and pasting sections from other job ads, sometimes even copying and pasting entire job ads. So you've got a role description and person specification that was likely to be wrong for the role it was originally written for, and is now even more likely to be wrong for the new role it's been copy and pasted for.

Another huge element missing from this approach is the context. Obviously a copy and paste approach generally ignores context altogether, but in agile, context is hugely important. How long has the organisation been using agile? How complex, fast-paced and uncertain is the organisation's external environment. In which internal contexts are they using agile? Is it just a software thing, or is it spreading into other areas too? What are the biggest problems with their use of agile that keep on cropping up? All of these context elements are hugely important, and are hugely useful to an agile practitioner in deciding whether they'd be a good fit for the role and vice-versa. Yet they are hardly ever included in job ads.

Moreover, agile people are typically problem solvers. They have to be, or they wouldn't want to work amongst uncertainty and complexity for a living. So if you give them fixed specifications for what the role will do and the type of person that should do it, you're pretty much telling them you're not interested in their problem solving, you just want them to do the things you think will solve your problems, by being the sort of person they want you to be. Honestly, you're not going to attract many great agile practitioners with a pitch like that.

So what's the solution to all of this? Well how about taking an agile approach to recruitment itself? One of the ways agile frameworks handle complexity and uncertainty is by doing away with writing specifications and requirements, and instead writing what are called user stories. These typically take the format of;

As a <type of person>
I want to <do something>
So that <I can achieve an outcome>

This way you describe who has the need for the thing, what need they have and why they have it. This user story format says nothing about what the solution is though. It only really describes the problem to solve. That way, you leave it up to the person working on the user story to use their knowledge, creativity and autonomy to solve the problem in the best way possible.

So rather than writing a person specification for a job ad, why not write down a series of user stories for the sort of things you're trying to achieve, then let applicants propose their own solutions to those stories when they apply, or during the interview? You'd get a much better sense of the candidates' knowledge and experiences, and you'd even get an interesting new bunch of ideas to boot. Rather than presenting the solution, present the problems, and ask people how they would propose to fix them. Isn't problem solving what you're hiring someone to do anyway?

The second part to this would be not just presenting the problems you hope the role will solve as a list of user stories, but presenting this list in a prioritised form too. Which of your problems are the most urgent, or the most difficult, or would offer the most value if solved? However you want to run your prioritisation is up to you, but put your top priorities at the top, and the lower priorities further down the list. That way candidates can propose solutions to the most important problems first, and provide simpler or less detailed solutions to the lesser problems, perhaps even miss out solutions to some problems altogether. You could even go a stage further and ask the candidates to give their own commentary on how you've prioritised the problems, and whether, based on their previous experiences, they might prioritise them in a different order.

Through taking these approaches, not only are you immediately more attractive to agile talent, you're demonstrating clearly that you're interested in two way communication and collaboration, that you've fully bought into and live the agile mindset. You're also allowing candidates to play to their strengths, creating an opportunity for you to learn new things from the agile community, and creating a whole bunch of interesting conversation starters to use at interview too.

Section 5: Working Environment

Idea 17. Agile Working Is Not Working From Home...
...Sort Of

> **Hypothesis**: Co-located or face to face working in agile environments is really about building high bandwidth communications between people working together, yet people are rarely if ever taught how to communicate well.
> **Experiment**: Own the delivery of, or even deliver directly, training and coaching in high-bandwidth communications for people and teams working in your lean-agile environment.

A few years ago I was hunting for a new job in the agile space. Every morning I'd get up, go through the job advertisement emails, and fire off applications to any that sounded interesting. One day, a pretty interesting job turned up, saying that a company needed someone to take ownership of implementing agile working across the whole organisation, right from the start.

Opportunities for genuine greenfield agile transformations, ones where there hasn't been a failed attempt already, are rare and combined with the idea that you'd have complete ownership to shape it in the way you thought best, it seemed like a pretty exciting job.

So after firing off the application, I gave them a call to follow up, which ended up being one of the oddest I'd experienced so far. It started well enough. They asked if I'd implemented agile working from scratch in organisations before, and I told them all about my successes in that area to date. After a bit of this though, they started asking more specific questions about how many desk space reductions I'd achieved, and whether I'd worked on multi-site desk space reductions. I had to admit that reducing desk spaces had never been top of my list for an agile transformation, but after a bit more discussion the confusion became clear.

The organisation had never heard of agile in the sense that we understand it. They thought it just meant flexible working in the very broadest sense, and so they had applied it to their current problem; the need to cut costs by reducing the amount of office space they were paying for. They wanted to do this by encouraging more employees to work from home instead, and because this meant that employees could work from both home and (sometimes) the office, they called this being agile.

The trouble was, our conversation wasn't just confusing, it was actually confused. I pointed out to them principle 6 of the Agile Manifesto;

"The most efficient and effective method of conveying information to and within a development team is face-to-face conversation." [7]

and suggested that perhaps getting people to work from home was potentially the *opposite* of agile working, as it might reduce collaboration between people.

However, I've since changed my ideas on this somewhat. Perhaps I've mellowed with age, perhaps I've had more experience of non-collocated working, but we have to remember that the agile manifesto was written in 2001. Back then, Internet access with any sort of decent speed was relatively rare for many people, and all we really had was email, fax, telephone or face-to-face conversation.

As a result, the underlying point of principle six of the agile manifesto is not just to make people be face to face, but to create what are called high bandwidth communications. You see, if you email someone, all you get are the words in black and white on a page. This is why, after about five emails backwards and forwards with someone,

---

[7] http://www.agilemanifesto.org/principles.html

you end up in an argument. You're lacking all the other elements that make up normal human communication. The tone of voice, the facial expressions, the body language, etc. The telephone makes things slightly better as you get tone of voice to some degree, but you're still lacking the facial expressions and body language. Stand face to face with someone though, and you get the lot, which is why it's called high bandwidth communication.

So for this reason, many people think agile teams and organisations need to have everyone collocated in the same place, meaning that that company's idea of 'agile working' was exactly 100% wrong. I'm not so sure though. Sure, face-to-face is always going to give you high bandwidth, but to fetishise it over other important factors seems to be missing the point.

First of all, it's not 2001 anymore, and technology has moved on. The growth of technology, along with a growth in Internet speeds, has meant that we have video conferencing that, whilst not always perfect, is certainly a whole lot better at creating high bandwidth communication than it used to be. Second though, just because people are face to face, this doesn't mean they're going to communicate well. Indeed for some people and some situations, putting people together can actually make things worse, either due to someone's lack of comfort at being around other people, or the fact that the people placed together just don't communicate with each other well.

This is where people management professionals can come in. First, obviously, to understand that the physical environment in which people work is an issue that is likely to be at the heart of a people manager's concerns. More important though, people managers need understand the point of collocated working in a lean-agile transformation, that it is about creating high bandwidth communication between people who are working together, and so ensure that these people are upskilled in a hugely important area; teaching people to communicate better, no matter where they are. Of course, some people managers may not feel this is necessarily their skillset to teach, but as the people who are often brought in to mediate when communications break down, isn't it in strongly in their interests to take ownership of ensuring that this training and coaching takes place? Then we could all move to truly agile working, no matter where we are located.

Idea 18. Create An Agile Workspace

> **Hypothesis**: Lean and agile workspaces give teams and individuals the autonomy and freedom to shape the space however they see fit. **Experiment**: Gather together all of your workspace policies into one place, and see how many of them you could reshape to give people the freedom to make their spaces their own.

If you work in the agile space for a while, you may come across the phrase 'cargo cult'. As always with these things, the truth of cargo cults throughout the ages in reality is quite nuanced, but the basic idea is that sometimes, people in different cultures misunderstand the relationship between a cause and an effect.

The story that best sums up cargo cults, whilst possibly more apocryphal than true, comes from the Second World War, where unsuspecting islanders in the Pacific Ocean suddenly became surrounded by a massive global conflict between hugely developed nations using technology they'd not seen before. As part of supporting the people caught up in the middle of this war, militaries on both sides of the conflict flew planes over the islands and dropped relief supplies for the islanders. If you'd not seen many aircraft before, especially not ones that dropped you food and supplies, then that sort of thing must have been pretty mind blowing.

Suddenly though, the war ended and the supplies stopped. The islanders of course wanted the supply planes to come back, so they built large replica aeroplanes and displayed them in prominent locations on the island, in the hope that this offering would entice the real supply dropping aeroplanes to return. Of course, they never did. However, this story then began to be used to sum up what is meant by the phrase 'cargo cult', inspired by the idea of islanders setting up a cult of semi-worship around wartime cargo planes. Through misunderstanding cause and effect, the islanders are meant to have thought that by recreating effigies of the planes, the real planes would return along with the benefits they provided.

Now, as I say, this specific story may not in reality be true, and there's a potentially unpleasant element of western imperialism around it in a way, but the phenomenon the story describes is most definitely true. In fact, it may well be happening in your organisation around lean and agile at this very moment, typically around how your organisation designs its physical workspace.

Typical workspaces in large organisations have generally remained unchanged for quite a while now. Almost like a factory, people have been expected to sit amongst rows and rows of desks, their exact location being determined by which team they are part of. They sit in front of their computer all day, occasionally heading somewhere else for a meeting, before returning to their desk. In fact, the only innovation in this model has been the introduction of hot desking, meaning you're not even guaranteed a desk any more, and you certainly can't leave anything on your desk from day to day, or customise your work area to a form that you would find most conducive to work.

However, over the last decade or so, large organisations have started to notice that the small little tech startups and digital companies they are now competing with have offices that don't look like this. They have more beanbags, and whiteboards, and break out areas, and are generally more colourful and fun. So as a result, often after visits to other companies, these large organisations decide they're going to make their workspaces more 'agile' or 'digital', and immediately put in an order for beanbags and whiteboards.

You may think I'm joking, but I'm really not. At one organisation I was working with a few years ago, when they found out that I used to work in tech startups, they asked me in all seriousness what ratio of bean bags to people tech startups usually provided. Was one bean bag per five people ok, or was in more like one for every three people?

The problem is that this entirely misses the point of what makes agile organisations agile, and is effectively creating a cargo cult around the whole thing. You can't just buy beanbags and whiteboards and hope that the gods of agile will bestow you with their numerous benefits and blessings. In fact, as so often with these sorts of things, the opposite might in reality be true.

On a simplistic level, the cargo cult here is operating because the physical object is removed from the reality of what the object was for. For example, when I worked in a small tech startup, we definitely did buy huge beanbags (eight feet long by six feet wide if I remember correctly) and leave in one part of the office for people to use. But we only did this because people were at times working so hard to keep the company afloat that they occasionally needed somewhere to go and have a quick sleep during the day. They also became great informal spaces for hanging out and having lunch. At the same time we'd probably have preferred to have had somewhere easier and better to get together and eat, we just couldn't afford to provide anything more than bean bags.

We also worked in an office that was a former warehouse with exposed bare bricks walls. Working in brick wall former warehouses is a common feature of tech startup companies, but the exposed brick walls didn't help us be more lean and agile. They were just a side effect of an office space that was far cheaper to rent than somewhere with nicely plastered walls. Old warehouses with bare brick walls can generally be rented more cheaply than nicely furbished offices, and in a small startup, every penny counts. However, I *genuinely* once saw a large corporate organisation consciously try to emulate this 'startup environment' by buying brick effect wallpaper and putting it up over the nicely plastered walls in their newly redesigned office. As if somehow it's making the walls look like exposed bricks that makes you agile.

There's a more important element to this cargo cult too, beyond just the silliness of buying brick effect wallpaper. You see, agile teams should have autonomy over how they do their work. It's all part of them being self-organising and self-managing. They also inspect and adapt continuously, seeing how things work, and making improvements to them based on what they see. So when an organisation decides centrally to redesign their workspaces to be more 'agile', usually spending a lot of money on it in the process, it often accidentally limits the teams' abilities to have autonomy and inspect and adapt how they work.

For example, one large corporate organisation I visited had the best intentions, and believed that agile teams needed wall space for creativity. Space for sticking sticky notes on and all that sort of stuff. So they purchased medium sized whiteboards, and stuck them up around the office in the middle of each of their big walls. Well intentioned as it was, this was actually the worst idea possible. The teams had previously been using the lovely large walls to stick their work on and promote transparency, but once a whiteboard had been stuck in the middle of the wall, what had been a large usable wall became just a much smaller usable whiteboard, giving the teams much less autonomy over their spaces.

Another organisation I met once got a company in to sell them 'innovation furniture', and bought in all sorts of brightly coloured and strangely shaped seating booths for people to have meetings in, then stuck them in the office's large open break out areas. Unfortunately though, due to the quirky design choices that the supplier had made, most of the booths could only seat two people, and the biggest of them could only seat four. So what had been lovely open spaces for whole teams of nine or so people to get together and collaborate were then made essentially unusable because of all of the furniture in the way.

The underlying theme with all of this is that you can't design an agile workspace from the centre, and often the more you spend on it the worse it will be. Agile teams like to have autonomy over how they arrange their spaces, with the ability to make changes as regularly as they need to. It's another reason why run down former warehouses are so great for this kind of thing, as the landlords of these buildings tend to be much less concerned and restrictive about sticking things to walls, or moving things around, or even knocking things down and rebuilding them differently.

If you truly want to create a workspace in which lean and agile can thrive, look instead to how many of your restrictive workspace policies you can remove, and how much you can give teams autonomy over how the shape their workspaces themselves. If you do insist on putting a budget behind creating an agile workspace, why not divide the budget up and give a chunk of it to each team, to spend on whatever they thing would be most useful? Copying the physical things you see in other companies will not make you agile, just as building replica cargo planes won't bring back the real supply planes. Understanding the reasons why tech company offices look the way they do is what really matters. Give teams the autonomy to decide how they need their workspaces to look, then empower them to turn that into a reality.

Idea 19. Enable Flow State Working

> **Hypothesis**: Some knowledge workers produce their best work by entering a flow state, free from distraction.
>
> **Experiment**: Help people identify if they might benefit from flow state working, and put in place the support they need to achieve this within your organisation.

How does your day at work normally go? Get in in the morning, check your emails and deal with the ones you can handle quickly? Then head into a day of back-to-back meetings and phone calls about various different topics, before going home in the evening feeling pretty frazzled, and not entirely sure what you've done or achieved during the day?

If any of this sounds familiar, then it's likely you're suffering from the effects of context shifting.

If you think the effects of context shifting aren't real, try this simple exercise, either by yourself or with your team. Put one minute on a timer, and write down as many times as you can the alphabet from A to Z. Count up how many letters you write in a minute.

Next put another minute on a timer, but this time, write down a letter from the alphabet and its corresponding number position in the alphabet. So, for example, A1, B2, C3, D4, etc. See how many letter and number pairs you write down in one minute. I can almost guarantee you that, even when you've adjusted for writing twice as much, you will have written far fewer things down in the second minute than in the first.

That's because in the first example, your brain is staying fixed on one context, writing the alphabet. Whilst in the second example, your brain has to continuously switch contexts between the alphabet and some numbers. The context shifting takes mental overhead itself, which slows you down.

How you believe you get anything done at all in a day where you're constantly shifting contexts between back-to-back meetings, interspersed with taking phone calls and answering emails, I've no idea.

But what's the alternative? Well, one alternative is to try something like the pomodoro technique. This is quite a popular one in agile circles. It involves setting a timer for a pre-determined amount of time, for example 25 minutes, and working on just one thing in that 25 minutes, without being distracted by anything else, before taking a break when the timer runs out. If you need to do more, set another timer for another 25 minutes, and so on. That way you can guarantee that you won't be context shifted for that time, and can in theory become much more productive as a result. Some people spin other ideas out of this, like only answering their emails at the beginning and end of each day, rather than being context shifted by them continuously throughout the day, or scheduling meetings only in the morning, leaving the afternoon for solo based work.

This works for some people certainly, but there's actually another approach that personally I find works even better, at least for me. Let me illustrate it with an example.

So far in my life, I've completed two masters degrees. The first one went well enough, not great but not appalling, but the second one went superbly well, and I graduated with a distinction level grade. What was the difference?

During the first one, I was still following the working style I'd been taught to follow at school, essentially similar to the pomodoro technique. Work for 45 minutes, take a 15-minute break, work for 45 minutes, take a 15-minute break, etc. I did it because I'd been told it worked, but I never felt entirely happy or comfortable doing it. In fact,

I always felt pretty stressed and distracted working that way.

By the time of the second masters though, I'd learned what my own personal knowledge work style is. It starts with me taking a few hours out just sitting quietly whilst playing a repetitive computer game, surfing the Internet or carrying out some other relatively mindless activity to slow my brain down, all whilst burning incense and listening to gentle ambient music. After a certain time, my brain would start to want to read research articles and start writing notes, or start collating and synthesizing the notes into an essay of some sort. I'd then lose myself in 24 to 36 hours of continuous reading and writing, broken only by answering the door to collect the takeaway I'd ordered online, and a bit of sleep in amongst it all too. As I was doing the masters at the same time as working, I'd generally start this process on a Friday evening, then emerge from it on a Sunday evening in time to iron some shirts and get my head back into the real world ready for work on Monday. The results were spectacularly good.

Later on, I learned that what I'd been doing wasn't just something that worked for me, it worked for other people too. It even had a name; 'Flow State Working'. Like anything in the field of psychology, it's just one set of ideas amongst various others, but a flow state is generally described as being intensely focused on the present moment and the task at hand. So much so that you lose yourself in it, gaining enjoyment from the sheer act of doing the task, to the point that you lose awareness of the passage of time and things going on around you. Personally I love getting into a flow state, losing myself in a piece of work and taking huge pleasure in making it as perfect as I can. Once I 'come round' from a flow state, and get feedback from others on what I've done, the results are usually superb too.

A flow state isn't easy to achieve though, and it generally can't be dropped into for just half an hour in between meetings. For me, I need a decent amount of time and space to get into it, and freedom from distractions too.

The research around flow state working also suggests some other interesting pre-conditions that are needed for someone to get into a flow state, most notably that the work provides a high degree of challenge, along with the perception that you have the high degree of skills to accomplish it. After all, if work doesn't present a challenge, then it's likely to provoke boredom, disinterest and a lack of attention and flow.

Conversely, if the work feels like a bigger challenge than you have the skills and knowledge to accomplish, then the work is likely to lead to anxiety, which isn't conducive to flow at all. Also interesting is the fact that research points to not everyone being able to enter a flow state of working, and that some people are better at it than others. Those who are good at it are thought to have autotelic personalities, that is, personalities that find reward from the intrinsic nature of just doing an activity, not expecting to receive any future benefit or external reward for the activity.

So how can people managers help in the pursuit of flow state working? To some degree, setting the conditions for flow state working is in the hands of everyone in the organisation. Managers especially need to recognise the potential value of it to their organisation, and give people the time and space to get into it when they need to.

However, there are definite blockers and barriers to flow state working that people managers could help to clear out of the way. For example, some people may find their flow state occurs more easily at home rather than in a busy office full of distractions, so relaxing your policies on home working could be an important element of facilitating flow states. Equally, helping to promote ideas like psychological safety in the workplace could be hugely valuable, given how toxic anxiety can be to flow. Some people of course may be better at achieving flow state than others, so there may be a role in helping people to identify their personality types and working style in order to understand if flow state might be for them. Finally, if flow state can only be achieved when a high degree of challenge is met with a high degree of knowledge and skills, allowing and supporting people to have the freedom to choose what they want to learn, and making that learning meaningfully available to them, could be a hugely useful contribution as well.

You may never have tried it. You may have fallen into it by accident and only now recognise it looking back at that time in your life. You may be as much of a fan of it as I am. Whatever the case, I firmly believe there is huge value to people and organisations to be gained through the further promotion of flow state working

Postscript: Readers of this idea may be amused to know that I started writing it on the 29th July 2019. Halfway through, my lovely wife asked me to check a website for the delivery status of a parcel we were expecting. My flow state thus broken, I didn't come back to finish writing it until the 31 July...

Idea 20. Scrap Your Leave Policy

> **Hypothesis**: Setting the number of days of leave people are allowed promotes the mindset that work is unsustainable, unenjoyable and a place where employees are controlled.
>
> **Experiment**: Scrap the maximum number of days of leave people are entitled to, and measure the impact on employee satisfaction, wellbeing and value delivery.

You might thing this idea is ridiculous, but bear with me. I suspect the reality is actually very different from how you imagine it might be.

When I started my lean-agile journey, I was working at a small digital agency in Bristol, UK. The agency had literally started life in somebody's bedroom, and slowly grown from there. Shortly after I joined, there were 8 of us crammed into a tiny office in a technology incubator building.

Given our humble beginnings and status, we'd not really bothered to do a lot of the HR policy work that large organisations typically do. One day we joked that we hadn't got a dress code for the office, so after bouncing some opinions around we wrote our new dress code into the company handbook. It just said 'Wear what you like, but be careful with purple'.

Another thing we hadn't bothered to codify was our leave policy either. It wasn't that employees couldn't take leave, of course they could. It's just that there was no formal leave allowance per year. If you wanted some time off, you agreed it with the directors, and just took it. I'm sure if it was something that was causing us a problem, we would have thought about it some more and written something down. However, we were all enjoying working there so much that we never really thought about it. Working at a startup for a cause you love doesn't feel like work, so as a result, we never really felt like taking much of a holiday.

Then, after a couple of years, we found ourselves on the brink of bigger things. We'd grown quite a lot, moved into a proper set of offices all on our own, and employed something like 20 people. On top of that, we were starting to pick up bigger and higher profile clients, so we needed to start acting like more of a professional, grown up company. One of the directors decided to hire someone in to help us do this, and empowered them to make any changes they thought right.

Straight away they drew up new more formal employment contracts for all of us, and one of the elements of the contract was to give us each a formal holiday allowance. A maximum number of days we could take as holiday, that expired and reset every year. We were a bit baffled by this, as of all of the potential issues we were now experiencing by growing rapidly, holiday allowances didn't seem to be one of them. However, what was even more baffling was what happened next.

Due to the time of year this was all happening, people noticed that their leave allowance for the year was going to be expiring and resetting relatively soon. So faced with a feeling that they were going to lose out on something they were entitled to, they started to book off large amounts of leave, In fact, so many of our developers began to book leave that we started to have serious worries about our ability to take on new work, or service existing clients if things started to go wrong.

Whilst we dealt with this short-term issue and balanced things out to keep us functioning and helping clients, over time we noticed that what we'd seen was actually a consistent pattern. When there was no leave allowance, and people could in theory take as much leave as they wanted, they took very little leave. Once we had a leave allowance, people took an awful lot more leave, as they now felt they had to use it all up.

Now as I said at the start of this idea, it may seem like a massive leap in the dark to abandon a leave allowance policy. It's also worth noting that in many countries there is a minimum number of days leave that employees are legally entitled to by right. There is though nothing stopping companies offering more leave than this total, in effect making the policy unlimited.

Interestingly, every time I've shared this story with others over the years, I've heard about the exact same pattern occurring in other organisations. People take more leave when you tell them how many days they can take. When you offer them unlimited leave, they seem to take less.

That said, the idea behind abandoning a formal leave allowance policy isn't to try to reduce the number of days of leave people take, although that can often be a side effect. It's actually to try to shift the mindset from the idea that work is a place you want and need to take leave from due to it being so exhausting or unpleasant. A great place to work means a place where people actively want to work, and can carry on doing so indefinitely without burning out. If you give people unlimited leave, and they start to take large amounts of it, there's your warning sign that people are unhappy or being made to work so hard that they're burning out. In effect, the number of leave days taken becomes an indicator for your organisational health, and by removing the cap on it, you remove a factor that might skew the reliability of that indicator's data. Focus on the root cause of why people want to take so much leave, rather than limiting their ability to do so.

This issue is also about decentralising decision making. By removing the leave cap, you're saying to employees that you trust them, and that you want to empower them to decide when they work and when they take time off, no matter when that may be. By removing leave from being an extrinsic, externally imposed motivation, and moving it to something each employee has autonomy over, you're taking another step to building a genuine lean-agile culture in your organisation.

Idea 21. Have A Contractor Contract

> **Hypothesis**: Bringing in contract staff can be useful in a transformation, but it can also cause significant frictions with permanent employees that can harm the transformation overall.
> **Experiment**: Consciously draw up and socialise a contract for what contractors can and cannot do, and their rights and responsibilities in comparison with permanent employees.

There's always an interesting skills shortage issue when undertaking a lean-agile transformation. If the skills you needed for lean and agile existed in your organisation already, you wouldn't need to run the transformation. Equally, this being a transformation, you probably want your existing people to be transformed, rather than having the cost and trouble of hiring a whole new workforce. As a result, you'll need to bring people in to transform your existing employees, but once the transformation's happened, you won't need to have those people around any more. So what's the easiest solution to all of this? Bring in short-term contractors.

This model has played out so many times now that amongst the agile community, I don't know of many people who are working in full time permanent positions as employees of a single organisation. Most people are contractors, and those that aren't get looked at oddly by those that are. Certainly bringing in contractors helps with the short-term need for lean and agile experience, but as with anything in this field, it's not something that should be done without some careful thought.

You see, so much of whether a transformation does or doesn't work depends on how it looks to the people on the ground in their everyday lives. If someone's been working loyally at an organisation for 10 years or so, they're naturally going to be suspicious of someone coming in all of a sudden and telling the organisation how to change. In effect the contractor can, unwittingly perhaps, make the permanent employee feel like they're been doing it wrong, or feel like they're being criticised unfairly, reducing the motivation of the permanent employee to engage with the transformation.

On top of this, there's the pay issue. Now lean and agile people are generally intrinsically motivated, that is, they're motivated by things like having autonomy over how they work, getting to be expert at what they do, and having a clear sense of purpose to work towards. In fact, if you read Dan Pink's book called Drive, his argument is that everyone is motivated in that way, as long as one factor is met. This factor being that you pay them enough that they stop worrying about money.

The problem here then comes in that contractors are typically paid a lot more than permanent employees. It's not their fault. They often contract because there's a shortage of their skills and they're in high demand, meaning they can charge a premium. They also don't get paid for their holidays or receive other work related benefits, and they are always at risk of having their contract brought to an end, so they have to build up a cash buffer to take them through the lean times too. However, a permanent employee earning a lot less doesn't really see or care about any of these things. What they see is someone turning up all of a sudden and earning a lot more than they are, even though they've not showed the loyalty and devotion to the organisation that the permanent employee has. By bringing in highly paid contractors, you risk bringing the extrinsic motivator of money back into people's consciousness, just at the time that you're meant to be moving the organisation to run along ever more intrinsic motivation lines.

Sometimes an unexpected side issue from this emerges too. Most organisations align their hierarchy grades with their pay grades. The more senior you are, the more you get paid. Sounds fair enough right? The problem is that when contractors come in, they actually reverse the way this runs. Suddenly it's not the more senior you are, the more you get paid. It's the more you get paid, the more senior you are. If the organisation still has hierarchy and pay as tightly coupled issues, then the highly paid contractors automatically have to get senior roles too. More senior than those who have been working at the organisation for a long time. How do you think that makes them feel?

Contractors are also people constantly living on the edge. If they're an independent contractor, then they need to keep demonstrating their worth in order to keep their income coming in. If they're working for a larger consultancy, they need to ensure that they keep the contract on behalf of the consultancy, and ideally keep expanding it too by bringing in more contractors. A 'land and expand' strategy as it's sometimes called. These motivations often cause them to do three things.

First, to take ownership of things and 'drive them forward'. If they can't do that, then at least to create the appearance that they're the ones owning and driving things, and third, when that fails, just take the credit for work that other people have been doing. After all, if it looks like the permanent employees can do this work by themselves, why would the contractors be needed any more? Of course, from what we've just seen around the 'pay equals seniority' issue, very often the contractors do have the hierarchical seniority to own, drive and take credit for things far better than permanent employees too.

So, consider this from the perspective of a permanent employee. You've been loyally working for an organisation for quite some time. Things haven't been perfect, but you've worked your hardest to keep the show on the road, and made personal sacrifices along the way, all in the belief that your hard work will be noticed and rewarded. Suddenly your organisation announces that it's 'going agile', and a whole bunch of contractors turn up. They know nothing about the organisation and what has and hasn't worked there before, but they're paid more than you are, they're sitting in more senior roles than you are, and they now want to take over and tell you what to do, unless what you're doing is going well, in which case they'll try to take the credit for it. What possible motivation do you have for joining in with the transformation?

I'm not saying that all of these phenomena happen all at the same time, often only some happen, or some happen more strongly than others. However, they're all perfectly possible, and over the last fifteen years of working in this space, I've seen each of them happen more than once, always with toxic results. So, what do you do about it?

My idea is to take a lean-agile approach to this problem. Make it transparent, create some policies to deal with it, and make those policies explicit to all sides. What you might call a 'contractor contract'. The contract could set out what contractors can and can't do, how roles and responsibilities between them and permanent employees would work, perhaps setting out a decision making framework that people can use to manage the relationships on a local level. For example, the presumption could be that the decision of a permanent employee will always trump that of a contractor, or at least set out clearly the circumstances in which it would do so. By facing these issues head on, setting out a framework for managing them, and making that transparent to everyone, so many of these needless issues could be avoided, and organisations, their permanent employees and any contractors they hire could all rub along much more harmoniously.

Idea 22. Run A Psychological Safety Audit

Hypothesis: A lack of psychological safety in the workplace kills innovation, experimentation and learning, and threats to it can emerge from anywhere.

Experiment: Run a psychological safety audit to find and remove the threats to psychological safety that your employees experience, and follow up with ongoing support.

One of the fundamental principles of a lean-agile transformation is that people need psychological safety. If you've not heard of it before, psychological safety is the idea that people are;

> *"…able to show and employ (their) self without fear of negative consequences of self-image, status or career."*
> (Kahn, 1990, p. 708)

In simple terms, feeling psychologically safe is the feeling that you can be yourself, that you can speak your mind freely, that you can do what you think is right, that you don't feel like you constantly need to ask for permission for things, and that you're not constantly looking over your shoulder for someone who is going to tell you off for something.

The consequences of a lack of psychological safety for lean and agile are severe. The whole point of lean and agile approaches is that people are free to learn and experiment, free to try things out in the name of continuous improvement. If they work, fantastic. If they don't work, then that's just as good, because that's now learning that can be shared with everyone else, so they can add that to their pool of knowledge when deciding what to try to learn from next. As Dan Pink puts it in his book 'Drive', people have to have a sense of autonomy, a sense that they are free to control their own direction and destiny, and know that the organisation will support them in doing this. Or, as Steve Jobs is claimed to have said;

> *"It doesn't make sense to hire smart people and tell them what to do; we hire smart people so they can tell us what to do."*

The problem is that, especially in a large organisation, there are a thousand forces out there, constantly trying to reduce people's psychological safety. They don't mean to, it's not a malicious thing, but there are many people, sometimes even whole departments, who don't realise that the actions they take are harming the psychological safety of others in the organisation. In fact, I often wonder at organisations that try to set up teams or programmes to promote innovation, without first of all tackling the teams and programmes that are busy telling people not to take risks, to look over their shoulder constantly, to fear doing something wrong to the point of being frozen into inaction. Let me give you some examples from my own time working in large enterprises.

Once, I was working in a large enterprise where we tried implementing lean and agile approaches around the production of their social media content. Just before Halloween one year, someone came up with the idea of photoshopping the brand logo so it looked like it had been carved into the side of a Halloween pumpkin. It was really professionally done, it looked superb, and it really chimed with the idea of the brand wishing people a happy Halloween. So we posted it on Twitter and Facebook, and the customer reaction was amazing. It generated the most likes, positive comments and shares we'd ever received, and overall the positivity coming back to us from the audience was something we'd never experienced before.

We were immensely pleased with how well our experiment had gone, so we shared the results with senior leadership, in the hope that others could benefit from the learning we'd created. Their reaction though was the complete opposite to what we had expected. Rather than giving us praise for the immense positivity in customer sentiment we'd created, an investigation was immediately launched into how this had gone so catastrophically wrong.

You see, the brand guidelines stated clearly that the brand logo could only ever be displayed in the colour black. By making the logo look as though it had been carved on a pumpkin, we'd made the logo look orange, and so broken that cardinal brand guideline rule. No-one got sacked, but it was a close run thing, and we were made to promise we'd never again try an experiment without first running it past multiple levels of sign off first, to check we weren't breaking any rules.

So you know what? We just never ran an experiment again. Why bother? The psychological safety to take risks and innovate had been taken away from us, no matter how positive the results were from our innovations.

Another time I was working in a large enterprise when I was asked to speak as part of a panel debate at an agile discussion evening that a colleague was organising. I happened to be running a training course in the same city, indeed the same company building, as the discussion evening on that date anyway, so of course I said yes. When the day came, I ran the course on the fifth floor of the building, then when it finished I went down to the ground floor, sat on a table at the front of the room and carried on doing what I'd been doing all day, answering questions about lean and agile.

Five months later I was abroad, and I woke up to find my work phone had been suddenly blowing up with missed calls, voicemails and text messages. Apparently the discussion panel I'd spoken on five months ago had also been advertised externally to the organisation I was working at, and unbeknownst to me, some non-employees of the organisation had been in the audience. So to speak at an event like that, I needed to have been officially licensed by the organisation, and I hadn't been, so I'd committed a breach of the organisation policies, and was now at risk of losing my annual performance related pay bonus as a result. Which seemed doubly annoying, as I'd spent the event promoting the virtues of the organisation as a great place for lean and agile people to come and work. Thankfully an escalation to someone senior got the issue dealt with, but again, you know what? From that day on I stopped bothering trying to attract good lean and agile people to the organisation by talking about the great work we were doing, in case I accidentally broke another organisational policy no-one had told me about.

These are just two examples, I'm sure there are many more out there, but the point of them is the same. So often, lean and agile coaches talk about psychological safety in the context of the team, or in the context of someone's immediate line manager. When in reality, the risks to psychological safety are often to be found coming from teams and policies that no-one seems to be aware of. It only takes getting caught out by one of these teams or policies once or twice though before someone gives up trying as a result.

The teams causing the problems of course weren't doing anything wrong. Like anyone else, they were trying to do the best job possible in promoting and enforcing the rules they had been given. It's just when they were given those rules, no-one stopped to think what their impact might be on the psychological safety of those in the wider organisation.

So what's the solution to this? Well, I think it's twofold, and both of the solutions sit with people managers.

The first is to run a psychological safety audit. Find a person or a team to come in as trusted, neutral and confidential colleagues, possibly people from an HR department, and ask people to report any of the things that stop them feeling like they have the freedom to experiment, to innovate and to learn when they're at work. The results may well be surprising, and it would perhaps be best if the teams that run the audit were then empowered to start tackling those problems, and the teams that are behind them. As I say, the teams causing the problems don't mean to do so, so HR professionals might be the best people to go in from a neutral standpoint and try to get the issue resolved.

The second solution is for these same auditing teams to then be an ongoing point of refuge for people's psychological safety. In the first example I gave, I had no one to turn to for help, and the event started me off on my path to leaving the organisation. In the second example, I was fortunate enough to have a fantastic colleague who stood beside me and kept me positive against the threat of taking away the money I use to feed my children, when I felt like I was being punished for helping to promote the organisation amongst the talent pool it wanted to attract.

How much better would it have been if the organisation had demonstrated its commitment to psychological safety by providing the contact details for someone who could help and intermediate in such situations? Someone you could turn to to say that your psychological safety feels at threat, and who would then come in to help resolve the situation, not with a specific organisational policy in mind, but with a focus on people's psychological safety.

Because once your people have lost their psychological safety, then they've lost their creativity, their innovation, their sense of autonomy and ultimately, their motivation.

Section 6: Performance Management And Reward

Idea 23. Establish Lean-Agile Performance Reviews

> **Hypothesis**: You performance review system may be actively driving the opposite behaviours required for lean and agile success
>
> **Experiment**: Experiment with performance management systems that create shorter feedback loops, promote collective accountability for success and support the self-direction and autonomy of the individual.

Once a year, many of us face our day of judgment. We've worked our hardest all year, and now we're called to account for our actions, being asked to explain why we're better than everyone else we work alongside every day. In effect, we're asked to justify our continued employment in the organisation. For some of us, this means to justify why the organisation should keep feeding us and our children.

You may say I'm being overly dramatic here, but I honestly don't think I am. Your annual performance review process is likely to be a hugely stressful time for your employees, and also potentially one of the most damaging things to your organisation's agility you can imagine. Here's why.

First of all, very often the criteria that an employee's performance is assessed against are set right at the start of the year. They're given their objectives, and expected to work towards them for the next 12 months. Or at least the next 9 months before their performance is assessed, given some performance management processes take at least three months to run. This is in essence big up front design, a classic sign of traditional project management. Agile organisations work on constant small feedback loops, doing a small amount of planning and a small amount of delivery to get some feedback on what to do next. Yet with an annual performance cycle, your feedback loops at best takes a whole year to complete.

More than this though, agile frameworks are about collective ownership. For example in Scrum, just like a team in the game of rugby from which it takes its name, the team is collectively accountable for its performance and the value it delivers. As such, Scrum doesn't even recognise many different roles in the team, most people are just called a 'development team member'.

So what effect do you think individual performance reviews have on collective delivery and collective accountability? Usually a very toxic one. For example, in your performance reviews, people are often graded against each other on a bell curve. For some people to do well, other people have to do badly. How does team collective ownership and delivery work when people know that their 'success' depends on others being seen to do badly? What incentive do employees have to help others, when helping others may help the others to rate higher on the bell curve than the person doing the helping?

As a small example of the odd results this bell curve approach can create, I was once told a genuine story of a team that contained a functional alcoholic. Someone who was in quite a bad way with their drinking, who started their day with vodka and sneaked out regularly for drinks whilst in the office. Despite this, the team they were in fiercely protected them, covered for them and made sure that they were never sacked. As even though this person did no work, they were worth keeping around, because when it came to performance review time, this person was guaranteed to get the bottom spot on the bell curve grading, meaning everyone else on the team did better. In a perverse way, the bell curve based performance review system incentivised keeping poor performers in the organisation.

On top of this, you want your employees and your teams to have autonomy as much as possible. Autonomy over how they work, how they organise, how they manage themselves and their teams. Yet once a year, your review process tells them that this autonomy is a nice idea, but now their manager is going to tell them how good they thought their autonomous actions were, and criticise the autonomous actions they didn't agree with. How do you expect your team's autonomy to be taken seriously after that?

Performance reviews are also toxic for another key element of agile adoption, the idea of servant leadership. There's lots to servant leadership, but the basic idea is that you devote your time to the service of others, helping them succeed, helping them look good, helping them perform better. Exactly the sort of role a Scrum Master plays in a team for example. Servant leaders are often crucial to a team's or an organisation's agile delivery, but when it comes to annual performance reviews, guess what happens? The servant leaders are told that they've had no visibility with the senior leaders who decide the performance grades (see idea 25), no-one has really seen what they've done (because they've been busy serving and helping others to look good), so they're easy pickings to occupy the 'underperforming' part of the bell curve that a certain amount of people have to occupy. As a result they leave, and the overall team and organisation performance goes backwards. Indeed, great servant leaders are generally only noticed when they leave, by which time it's too late.

On top of this, there's a massive potential cultural problem here too. If you're adopting lean and agile approaches, you can spend all year working towards them, but unless you write them into the performance goals for employees, there will always be an incentive for people to ignore them in favour of whatever the performance goals say. Even if the goals do say the right things, are the managers running the reviews and calibrating the gradings fully aware of what lean and agile mean, and the sorts of new behaviours that should be being rewarded? Very likely not. As a result, it's perfectly possible that you can spend all year promoting a transformation, only for much of your hard work to be completely undone at the end of the year.

So with all of these problems with performance reviews, what's the alternative? Well, there are many alternatives, and many forward-looking organisations have already adopted some or all of them.

The first is just to abolish the annual nature of performance reviews, and instead move performance management to continual feedback conversations between managers and employees. This certainly ticks off the first problem we noted around the traditional annual system being all about big up front design. However, to my mind it still doesn't really tackle the other problems of autonomy, and collective accountability. Nor does it address the issue of traditionally minded managers reviewing performance against the old criteria in their head, rather than the new criteria the organisation is aiming for. In a way, you actually risk making this problem worse, by increasing the number of opportunities they have for encouraging employees back into old behaviours.

The second is to move to a system of collective performance reviewing. If the team has autonomy over its ways of working, shouldn't it also have some autonomy of how those ways of working are assessed? Shouldn't team members be encouraged to run performance reviews for each other? You could even extend this idea to situations where bonuses are attached to performance reviews. Rather than managers allocating bonuses to individuals based on the manager's opinion, why not move to a system where teams as a whole are given bonuses, and the teams decide collectively how to divide that bonus up amongst their members?

Best of all though, to my mind at least, is to abandon this idea of performance reviews altogether, and instead spend the time and money supporting and coaching people to improve their own performance. Show them that you're an organisation that believes in people's intrinsic rather than extrinsic motivations. That you believe people inherently want to grow, develop and improve themselves, and that all the organisation needs to do is provide servant leadership in order to enable them to do so, giving them the tools, resources and opportunities to enable them to review their own performance, and take ownership of their own growth. What this would look like in different organisations would probably differ quite widely, but I'm certain it would be hugely powerful. If you know of interesting ways organisations have implemented this sort of approach, I'd love to hear about them.

Idea 24. Teach Receiving Feedback, Not Giving It

Hypothesis: Systems that shape how people give feedback often limit it. There may be greater value in teaching people how to receive feedback instead.

Experiment: Identify what people fear or find difficult about receiving feedback, and coach them to overcome it.

Pretty much any agile framework you want to look at will have a focus on transparency. A focus on seeing things as they really are and, as a result, basing decisions on empirical evidence rather than opinion or hope. This sounds like a pretty simple idea that's hard to disagree with, but in reality it's actually very difficult to implement.

As human beings, we often struggle with transparency, especially when we've been brought up by the education system to see failure as a negative to hide away from, rather than a learning point to embrace. As a result, when things sometimes go wrong, as they inevitably do, we try to put a positive spin on them, rather than being transparent about the failure and looking to fix the root cause of the problem. This ingrained human difficulty with transparency is hard to shift. However, there is an aspect of it that might be easier to change, that being the issue of feedback.

Feedback is another integral element of lean and agile, and often goes hand in hand with transparency. Making things transparent is necessary in order to create continual learning and improvement, but in and of itself it is not sufficient. We make things transparent in order to give feedback on them, and it is that feedback we use to improve what we do next.

Now lots of organisations these days have feedback built into their performance management systems for employees. Generally this is an annual process, giving employees feedback on their performance throughout the previous year, and guidance on what their performance should look like for the coming year. Sometimes it's more regular, or sometimes a lighter touch more regular approach goes hand in hand with an annual cycle of feedback too. We cover these issues more in idea 23, but there's another side to this feedback issue that's less often considered; the idea of there being an approved organisational approach to giving people feedback.

I first came across this issue when I was working in an organisation that had done exactly this, and set out a model for how to give feedback. Presumably for the best of reasons, possibly due to recognising that people found giving and receiving feedback difficult, they had set out a model called 'Even better if…'. This approach basically said that if you wanted to give someone feedback, you shouldn't talk about what went wrong, or what they did badly, you should reframe the feedback on the situation as something that would have been 'Even better if…' the person had done something different. So for example;

> *"I note what you did in situation X, and it would have been even better if you had done Y."*

The thing is, well intentioned as it may have been, there were actually a number of significant problems that this approached caused.

The first was one of one simple grammatical confusion. In many instances, the new format for feedback didn't actually change how people gave feedback at all, it just made them write the words 'Even better if' at the start of their complaint. As a result, sometimes feedback just made no sense. For example, if someone emailed a strong complaint about someone or something they had done, it didn't make sense to say doing something else would have been the 'even better if', as the situation they were describing was not good in the first place. A bad situation can be made better, but technically only a good situation can be made *even* better. This may seem like a semantic point, but it's pretty important. If you receive feedback that makes no sense grammatically, you're less likely to believe it.

What's actually going on behind this grammatical point is a bigger issue though. Ultimately people want to give the feedback they want to give. If having to follow the approved format means the feedback no longer makes grammatical sense then so be it. They'll follow the format and still say whatever they were going to say in the first place. I suspect this point is actually pretty powerful and universal. You can give people all the formats and processes for feedback that you want, but if there's something that they want to get off their chest, then they will get it off their chest regardless of your format. In essence, your format becomes a game they have to play to do what they would have done anyway.

I say that, but there might in some instances be an even worse unintended consequence of a prescribed feedback format that's taken too far. What if the format is so prescribed, stating what can and cannot be said, that it actually shuts people down from giving feedback? If what you want to say contravenes the tools you've been given for saying it, then might it be easier not to say it at all? This would be a horrible unintended consequence of a policy that was no doubt intended to help rather than hinder. One that threatens the transparency and feedback culture so essential for allowing lean-agile approaches to thrive.

Once a culture of what can and cannot be said in an organisation becomes entrenched, it can be very hard to shift. I'm an agile person, and have been for years, but I notice even I get caught up in cultures like this if I stay in them too long. My warning sign is typically when I write more emails than I send, writing emails that I then have to delete because the culture has got inside my thought processes and started causing me to self-censor and reduce my transparency.

So what can we do about this? Well the obvious thing would be to encourage open and honest feedback, removing any attempts to define how feedback should be given to others. But that potentially puts us back in the same problem that people were trying to fix in the first place, that people find it the whole process of feedback difficult and sometimes troubling. If you just make it a free for all, then chances are you will get issues emerging as people struggle with the concept of transparency.

So if you actively need to make the feedback process easy for people, but can't tell them how to give it, what are you left with? Well, rather than mandating how feedback is given, how about coaching people on how to *receive* feedback. How this would look might need to vary dependent on the person and the context, but it would solve a large number of problems. At once, you're no longer constraining how people give feedback. Indeed, you're freeing them up to give as much feedback as they want in pretty much any way they want, because the emphasis has been placed on the receiver to enable them to receive it more easily.

At the heart of this approach is likely to be an element of systems thinking called 'personal mastery'. I shan't go into systems thinking too much in this book, as it's a huge topic that would take us well away from the field of people management. However, personal mastery in systems thinking is described by Dantar Oosterwal in his book 'The Lean Machine' as;

> *"a special kind of proficiency and confidence in one's ability to the extent that one is able to question it openly and accept criticism from others."*
> (Oosterwal, D., 2010, p53)

In short, you create people who are so constantly open to learning and feedback that they welcome it as the valuable thing it can be, rather than fearing it as an attack on their very nature.

This work would likely go along with broader coaching on communication skills, so often one of the big missing pieces for lean-agile adoption, along with creating a culture of psychological safety around failure. Another element of it might be to encourage people to apply critical thinking skills around the feedback they receive, enabling them to take ownership of how the feedback will or will not be used.

In this way we'd start to remove the fear and rejection people experience through feedback, and also empower them to act on the feedback as they saw fit, again decentralising decision making to the level of the individual.

As I say, the details of how this model would look in your context would be up to you. But as a mindset shift, I really do think that we enable transparency, continual growth and learning far better by helping people to receive feedback rather than trying to guide those giving it.

Idea 25. Separate Performance Appraisals From 'Visibility'

Hypothesis: In a performance management system run on people making themselves visible, your best lean and agile talent will often perform poorly and progress slowly.
Experiment: Encourage senior people to walk the floor and visit people and teams to see work first hand, so that performance appraisal does not rely on people making themselves visible to the senior person.

I have to admit something right at the start. This idea is one that I have a strong personal bias around, in as much as the whole issue genuinely annoys me.

It's summed up by a job advertisement I once saw shared on LinkedIn by someone I used to work with. They shared the advert and added their own personal comment on the role, stating that one of its biggest benefits was that it had 'great senior stakeholder visibility'. Not that it was a great team to work in, or that it was a great role for personal growth and developing new skills, or even that it was a key role that delivered significant value to the organisation. Just that it was a role that would get you noticed by senior people, with the implication being that that's the thing that will lead to you getting ahead in the organisation.

I've seen this idea come round again and again. Unless you get time in front of senior people, and get them to see what you do, then you won't get a good score in the annual calibration session for your performance review, you won't get a good bonus and you won't get promoted. I once even knew someone who was treating their personal 'senior stakeholder visibility' like a project in itself, creating a spreadsheet to track every senior person in the department, and creating a plan for how they'd get some of their work in front of each of them individually at least every three months. Try applying the 'would a customer be happy to pay for that?' test to that type of employee activity. From a customer perspective, the whole visibility thing falls squarely into the 'unnecessary waste' category

Now the problem is that when it comes to employee recognition and career progression, there probably does have to be some element of senior person involvement in the decisions that are made. There are very, very good arguments to be made for the idea that organisations should flatten their hierarchies, and that career progression should be through collective agreement, but the reality is that few organisations have reached that sort of place yet, and for many, such things may be a leap too far initially. So we need to address the problem as it is, not as we wish it would be. The question then is, how do senior people get involved in employee performance and career progression approaches in a way that doesn't accidentally incentivise employees to spend more time on 'senior stakeholder visibility' than they do on delivering value to the organisation itself?

Of course, the reality is that senior people are only human. They have the same number of hours in the day as all the rest of us, and probably far too much work to do in the limited time that they have. No wonder they've fallen into a pattern of 'out of sight, out of mind' when it comes to identifying talent in their organisation. It's not their fault specifically, it's a product of the system they work in. Besides, why even should they care about the career progression of each member of staff? Shouldn't it be up to the person wanting to progress to demonstrate why they should be promoted? Surely you want the most motivated people to progress to the higher ranks, as that motivation will be useful in the future?

To a degree of course yes. However, isn't it in the senior person's best interest to ensure that the people who are getting promoted are the ones who are the best at delivering value to the customer, not the most motivated at getting themselves promoted? Don't you want to promote the people who help promote the goals of their team, their department and their wider organisation, not their own personal goals?

From a lean and agile perspective, you really do. As W. Edwards Deming noted time and again, it's getting the organisation to work like one co-operative system that matters, and yet so often people look to optimise themselves and their immediate surroundings. The thing is, lean and agile people generally know this, and as a result they often don't care about their personal rating or the perceptions people have of their performance. They see their role as being to help, support and coach others to be the very best they can be, all whilst ensuring that it is the wider system that is being optimised. It's why I say your lean and agile talent will never be strong performers. How can they be, when many of them will agree with Lao-Tzu's aphorism that;

> *"A leader is best when people barely know that he exists, not so good when people obey and acclaim him, worst when they despise him.*
> *Fail to honour people, they fail to honour you.*
> *But of a good leader, who talks little, when his work is done, his aims fulfilled, they will all say, 'We did this ourselves.'"*

As a result, in a lean and agile environment, it's perfectly possible that your performance management system will be actively penalising the people who do the most to make your organisation lean and agile.

So how do we fix this problem? Interestingly, lean and agile may give us the answer.

The answer, I believe, lies in a simple behaviour shift by the senior people. If senior people still need to be involved in the performance and promotion decisions, and they have limited time to understand who is doing what, then one of the best ways for them to understand what's going on in their organisation is by going and seeing. Lean calls this 'genchii genbutsu', commonly described as 'go and see'. Go to the real place where the work is happening, and understand first hand what the reality is.

This doesn't need to be a difficult process. Just walking around the office for a while every day and stopping to talk to people can reveal huge amounts about what's really going on on the ground, and give senior stakeholders far greater amounts of high quality information on which to base their decisions. Especially when compared with sitting at a desk being fed reports by middle managers, who have a vested interest in making it seem like everything is going great.

Not only will senior people themselves benefit from the act of walking around and seeing what's going on, by doing so, they remove the need for people to spend time focusing on senior stakeholder visibility, as now the senior stakeholders are the ones creating the visibility, leaving the people lower down free to spend time on the things that really matter, like delivering value to the organisation. In fact, it's now in the lower down employee's interests to make sure that most of their time is being spent on value delivery, as they will never know when someone senior may stop by and ask how things are going.

This may sound like an impossibility in your context, but my experience so often has been that it only seems like an impossibility because of the way things currently are. Walking around from time to time, or taking a desk sitting in amongst a team is not a big shift in effort on behalf of the senior person, all they need to do is trade off the time for this against some of the time in their diary that is spent on a similar but less effective version of the same activity. For example, if they cut one or two 'status meetings' or 'steering committees' out of their diary, and replaced them with some visits to the actual teams who were having their status reported, I suspect they'd have a good amount of time for each team in turn over the space of say three months or so.

For all this is a small shift in effort though, it might lead to a big increase in productivity, as employees stop focusing on getting noticed, and start focusing on delivering value instead. At the same time, senior people start to get a real picture of what's going on in the organisation, rather than a picture that's been passed through a layer of formal reporting and balanced scorecards, enabling them to make even better decisions. On top of that, your lean and agile talent may even start getting the recognition they don't crave, and the value they create may begin to be truly recognised. Everyone's a winner.

Idea 26. Act As If No-One Comes To Work
To Do A Bad Job

> **Hypothesis**: Everyone in your organisation comes to work every morning to do the best they possibly can. If it seems like they're not doing, the problem may be with the wider system they work within. **Experiment**: Approach every HR grievance, every disciplinary action, as if the person is not to blame, but that they might be a victim of how the organisational system currently operates.

It's funny isn't it? So much of our current performance management system is centered around the individual, and how likely they are to fail. If something goes wrong, it must be someone's fault. If something takes longer to deliver than was expected, someone must be being lazy, or have dropped the ball. If you let people work at home, they'll obviously just stop working and be mess around. If someone does block, frustrate or annoy us, it must be because they're personally not a nice individual.

The problem with these views is that most of the time, they're largely nonsense.

I have a test I run in the training classes I teach. I ask people to put their hands up if they've ever got up in the morning and thought:

> *"Today I'm going to go into work and annoy people. I'm going to block whatever it is they want to be done, question their decisions, frustrate them and generally be as difficult to work with as possible."*

To this day, out of thousands of people I've trained, only one person has put their hand up. When I asked why, they said that they hated their job and had already handed in their notice, as they didn't like the person their job had turned them into. Fair enough really.

Within what they said though lies the heart of this issue. They felt that their job had turned them into the kind of person that wanted to cause problems for other people, the sort of person they didn't want to be. The thing is, as W. Edwards Deming put it;

> *"I should estimate that in my experience most troubles and most possibilities for improvement add up to the proportions something like this: 94% belongs to the system (responsibility of management), 6% special."*
> (Deming, 2000, p315).

In that statement, he was saying that people actually want to work hard, they want to take pride in what they do, they want to succeed. No one turns up to work to do a bad job. When we do experience failures or delays, very often they're caused by the misalignment of people, or incorrect assumptions, or personal or team goals that conflict with other people's goals within the same organisation. Problems caused by the system people are working in. A system designed by managers, full of what are sometimes called 'local optimisation thinking mistakes'.

If someone's getting in your way, are they intentionally doing a bad job to frustrate you, or are they just doing the best job they possibly can in a way that unfortunately conflicts with the job you want to get done? The chances are that it's the latter, but as human beings we tend to think that it's the former.

Take the idea of DevOps for example. An entire movement founded on the problem that software developers want to keep putting new code into production as often as possible, but operations people are being expected to keep that code stable and secure when it is in production, being used by customers. So the last thing operations people want is developers putting new code in there, it might break something. Which is why we now have DevOps, trying to create alignment between developers and operations people, or between Dev and Ops. Neither side has been doing a bad job, they've both been doing a brilliant job. It's just that their jobs have been set up with competing goals by the wider system in which they operate.

To look at some other examples, if something's going slowly, are there greater gains to be made in making the person work 10% faster, or by reducing all the weeks of delays they experience whilst waiting for senior management sign off on their work? Chances are it's the latter, but naturally senior management always tends to blame the former. If someone works from home, are they actually working just as hard as they always do, or do you just imagine they're doing nothing because you can't see what they're doing? Changes are it's the former, but that you imagine the latter.

My personal anecdotes and experiences aside, there's actually a whole school of academic thought around this issue too. It's called Theory X and Theory Y. This theory was developed in the 1960's by Douglas McGregor at the MIT Sloan School of Management, and has to do with how you see people in the workplace. In essence, it's all about whether you see people as self-motivated and able to succeed without management (Theory Y) or whether people are inherently lazy, money motivated and in need of active management and supervision to get anything done (Theory X).

Now this isn't a theory just about how humans interact with each other. It's actually a theory about different ways of running teams and organisations in different circumstances. In Theory X, workers are seen to have little ambition, little ability to motivate themselves towards work and less intelligence than those in more senior positions. To deal with this, managers feel like they need to set individual goals for people, track their work and enforce delivery through carrot and stick. It may not sound like a fun or pleasant approach to the workplace, but it does produce results. This close attention, individual goal setting and intensive performance management tends to produce standardised processes, uniform results and high degrees of employee specialisation in tasks. If you're running a production line, where someone just needs to do their job, and not worry about the wider system, then these are just the sorts of things you want.

Theory Y on the other hand believes that workers are self-motivated. That they're keen to work hard even if no-one is telling them to do so. That they're not especially motivated by money, and that they're keen to better themselves and their work through self-reflection. It's pretty much the opposite of Theory X.

I think you can tell where this going. If Theory X is good for standardised work on a production line, Theory Y is much better for knowledge work, where the work is ever changing and unpredictable, where people need to self-organise, continuously inspecting and adapting their work and ways of working. In short, whilst Theory X is good for traditional ways of working, Theory Y is what you need for lean and agile.

The problem is, so many people have been brought up on Theory X that getting them to adopt Theory Y can be really tricky. If you think about it, much of our education system is based on Theory X. Teachers set goals for pupils, reward them with good grades and prizes if they do well, and punish them if they transgress. I suspect it's the same reason some people who do well at school find that they struggle at university, having gone from an environment where they did what they were told to do for fear of punishment, to one where they're expected to motivate themselves to learn, grow and develop.

To compound this, when we finally leave education, we often start our careers at the bottom of an organisational hierarchy where we're more likely to be exposed to Theory X styles of management. So it's no wonder that by the time people get to management level, they're fully on board with Theory X, and see attempts to move to the Theory Y approach required for agility as naive hippy nonsense.

Quite apart from its poor fit with the self-organisation and intrinsic motivation required within agile teams, there are two bigger problems that Theory X approaches cause for agility.

The first is that Theory X very often becomes self-fulfilling. If you approach self-motivated knowledge workers with a Theory X stance, there will often be a short period of storming, before the worker either just walks out and leaves, or gives up and submits. In either event, Theory X becomes self-fulfilling prophecy. If the worker walks out and leaves, then it's easy to think that the whole idea of self-motivation is not true, as if the person were self-motivated they would have stayed. If they submit, then Theory X looks more and more necessary too. The worker no longer has any intrinsic motivation (you've demoralised and beaten it out of them) so clearly they need closer and firmer management to achieve results.

The biggest problem to my mind with Theory X though is the one we looked at at the start; the way it blinds you to systemic issues. So many of the problems organisations encounter are systemic. Individuals who are performance assessed on an individual basis, teams working in isolation to each other, people optimising things to suit their local situation regardless of the effect on the wider organisation. The problems that these practices create happen again and again, but if you've got a Theory X mindset, then it's easy to explain them away as the failings of individual people. If a team produces the wrong thing, blame the manager. If the quality of a team's output starts to suffer, blame their tester. Whatever happens, Theory X makes you think it's always somebody's fault, and never the fault of the way the entire system is designed or way the organisation is structured.

A lean-agile approach means looking at the whole system, building collaboration, inverting organisational hierarchies through servant leadership and so much more. If you've got a Theory X mindset, you're very unlikely to do anything of these things, very unlikely even to see the purpose or value in doing them. Agile people are much more firmly in the Theory Y camp though, and if you want to adopt an agile mindset, it's a camp you need to move over into as well.

Idea 27. Stop Your Promotion Strategy Harming Your Value Delivery

> **Hypothesis**: Not everyone in your organisation helps it to achieve its primary purpose; the creation of value.
>
> **Experiment**: Run an analysis of where value is truly being created in an organisation, and refocus your supportive efforts on those areas.

Here's an interesting problem. What's the main purpose of your organisation? I don't mean what products do you sell or what services do you provide. I mean what is the ultimate point of what your organisation does? Chances are, it's to create and deliver value. Value to your customers or service users, and in return, value back to the organisation itself.

So, who in your organisation delivers value to your customers? Everyone, right? Certainly that's what you'd want if the purpose of your organisation is value creation. Why would you employ people who don't help the organisation to achieve its purpose?

The thing is, very often this might not actually be the case. There are different schools of thought on this issue, but they all seem to agree on the one thing. Some people in your organisation deliver value more than others. Some people potentially deliver no value at all.

If you look at something like the Toyota Production System, a big foundation for lean thinking, they categorise various different types of waste in an organisation. Waste being things that don't add value, basically things the customer won't pay you for. There are different types of waste, but you have to ask, which people in your organisation are most responsible for causing them?

When it comes to waste, Toyota talks about muda, muri and mura, meaning non-value adding activities (waste), overburden and unevenness respectively. Muda (waste) then breaks down into seven different types of waste as well, many of which I'm sure you'll recognise.

For example there's;

> *"Inventory — a capital outlay that if not processed immediately produces no income; Waiting — products that are not in transport or being processed (and) Over-processing — when more work is done than necessary, or when tools are more complex, precise or expensive than necessary"* [8]

Add into those ideas the idea of muri, where people or systems are being overburdened with work, and mura, which is often the unevenness in the flow of work caused by poor organisational systemic design.

So if these wastes are something to be avoided, if they are things that hinder the creation of value, what causes them? What types of people in an organisation would have the ability to delay things, either through asking work to be paused whilst they review it, or by designing organisational systems that don't focus on reducing delays? What types of people would have the ability to design processes that are more complex than necessary? What types of people have the ability to cause others to be overburdened with work?

The unfortunate elephant in the room is that very often it's the more senior people in an organisation, the leadership, who are directly or indirectly causing these wastes, and slowing down or reducing the creation and delivery of value. After all, who else has the primary authority and ability to cause these wastes to happen?

Now this is not to say we need to do away with senior people and leadership within organisations at all. Just because this level of people *can* cause these issues doesn't at all mean that they do do so. However,

---

[8] https://blog.toyota.co.uk/muda-muri-mura-toyota-production-system

we have to be open to the idea that this might be what's happening, and acknowledge the harm that this can cause to an organisation's primary purpose of delivering value.

But what has this got to do with people management? Surely the perception that senior leaders slow things down and create waste, wittingly or unwittingly, is ages old, and is a problem for the organisation more widely?

Well, yes and no. Sure, it's a problem for the wider organisation, but people managers might unwittingly be adding to this problem in various ways.

Assume for a minute, for the purposes of argument, that people at the lower levels in an organisation do most of the work that creates the value, and people at higher levels are at the greatest risk of delaying or preventing the creation of value. If this is the case, then surely you want to focus more on supporting the people at the lower levels rather than the higher levels in a number of different ways?

First, when it comes to training and skills development, doesn't it make more sense to invest in the lower level workers than the higher-level ones? If they're creating most of the value, you want them to be able to do it to the absolute best of their abilities. So why then in many organisations do people at the lower levels find it so hard to get budget and approval to get training, whilst people at higher levels can even get whole university level courses funded in the form of company sponsored MBAs?

If you look at remuneration too, surely it makes sense to share the profits most of all with the people who are creating the most value? Why then do organisations typically pay far more to people at the top than people at the bottom? On a purely economic analysis, this phenomenon may be completely backwards.

Above all else though, why do organisations spend so much time designing schemes to enable people to move from lower levels to higher levels in the organisation? From graduate recruitment schemes fast tracking people to the top, to mentoring schemes with senior leaders designed to 'spot future talent' and many more such schemes besides, there's a real risk that the organisation is unwittingly pulling its best people away from the most valuable work that needs doing, and moving them to areas that create much less value. Sometimes pulling them into areas that can even prevent value from being created at all.

Now I'm very definitely presenting these ideas as hypotheticals rather than absolutes. There are senior leaders in organisations who very definitely help with value delivery. There are doubtless people at lower levels who don't deliver much value, for a whole variety of different reasons. It is though something to bear in mind when looking at how to bring lean-agile thinking into an organisation, that some of the current training, reward and promotion structures in an organisation might unintentionally be reducing the organisation's ability to create value, and thus reducing its ability to meet its primary purpose.

Idea 28. Base Agile Salaries On Value Delivery

> **Hypothesis**: Your salary progression policy may be based on an old world of linear skills and competencies progression within a predictable environment, some or all of which now no longer apply. **Experiment**: Examine your competency framework against the skills and competencies that are now most valuable to your organisation, and re-benchmark them against salaries being offered in the external marketplace.

A few years ago I got into a bit of a discussion with a recruiter. The job they were trying to hire for looked pretty interesting, and it was a senior level post, with a lot of influence and so a lot of responsibility to go with it. In the process of the initial discussions, we ran into a problem.

The discussions were going well, but then we got onto the issue of salary. They asked what I was earning at that time, and I asked what the relevance of that question was. It was a senior level job with significant responsibility, and I knew roughly what the market rate for that job would be. Besides, I was contracting at the time, working to build up a company, so what my salary was wasn't as easy as just showing someone my monthly pay slip.

However, the recruiter thought that this was an issue. The company the role was being hired for had a strict policy that they would only pay new hires 10% more than their current or last recorded salary, and that wasn't going to change. We got into a bit of a debate about how this didn't really work from an agile perspective, which I hope they found interesting, but ultimately the conversation fizzled out.

The chances are though, the same issues I was discussing with that recruiter are ones you may encounter in your own organisation too.

You see, things like salary and benefits can have pretty big impacts on the organisation's profitability, so understandably they're closely controlled and often determined through centrally decided policies. Which is how you can end up with a blanket 'no more than 10% salary increase' rule. There are genuinely though a number of serious issues this sort of approach cause for a lean and agile transformation.

Fundamentally, agile is all about the delivery of value. The first principle of the agile manifesto states;

> *"Our highest priority is to satisfy the customer through early and continuous delivery of valuable software."* [9]

As such, any role in an organisation will have, or at least should have, a business value attached to it, as every role should be about delivering value. How you calculate the value of a role is usually a combination of different factors like delivery, cost, risk reduction, knowledge sharing, innovation and anything else that supports or relates to the value your organisation creates. As such, what someone in a role is paid for, or at least should be being paid for, is a share of the money generated by the value they create. You see this most clearly in sales roles, where this idea is used as an incentive, with sales people being paid an agreed percentage of any revenue that their sales generate. As a principle, it should be true of any role though, although with an obvious tension between the organisation and the employee around what percentages of the value generated each side gets to claim.

All of this said though, it is clear that the business value for a role, whilst related to many different factors, is completely unrelated to the previous or current earnings of any candidate applying for the role. Saying that it is related makes literally no sense, as what someone earned somewhere else has nothing to do with the value they will create in their

---

[9] Incidentally, this may say software, because it was written as 'The Manifesto For Agile Software Development', but if you take out the word 'software' and replace it with 'products' or 'outputs' or 'deliverables', or even just 'stuff', then the values and principles in the manifesto still work, and can be applied to pretty much any field of work that involves complexity and uncertainty.

new role. Imagine going into a shop and buying an item for $1, then going back into the same shop later to buy a different item, which involved different effort and complexity to create, but insisting that you will only pay $1.10 for it, as you can only pay 10% more than what the previous item cost. The value of the two items may be completely different, just as the value of the two roles may be completely different, so you can't apply a linear cost progression between them.

Now of course you may argue that if someone is moving steadily up the career ladder, there would be a linear progression in the value they create, and so there should be an associated linear progression in the salary they are paid. In some cases this will of course be true, but to then enshrine this idea into a standardised policy across an organisation entirely misses the point. Assuming a linear progression of skills and experience only really works in predictable and repeatable work environments. Places where you can know what the starting state, intermediate states and ultimate end state of a career path may look like. But when you're working in lean and agile, you're saying that things are now complex, uncertain, and fundamentally unknowable, so this whole idea breaks down. As we look at in idea 16, in lean and agile, even the idea of defining a detailed job description up front for a role may not makes sense. So if the role is unknowable, there can't be any known linear progression for it, meaning a linear progression of salary, rising 10% each time, doesn't make sense.

A 10% progression for a role also ignores the ability of people to grow and develop their skills faster than an organisation may typically expect. What if someone takes some time in evenings and weekends to increase significantly their knowledge and aptitude in a field, perhaps even investing significant amounts of their own resources in doing so? A greater than 10% increase in their ability may well lead to a greater than 10% increase in the value they can deliver, so should they not be rewarded with a greater than 10% increase in their salary? You often see this sort of situation when someone has been working at the level of a single team for a while, building up lots of skills and experience, then brings all of this together in some sort of vocation based external qualification, such as an MBA. They now have the knowledge and experience to operate at a much higher level, but due to their low 'team level' salary up until that point, they are unable to be sufficiently rewarded for their efforts. Some people I've known this happen to have headed off to do contract work, albeit reluctantly, as they had wanted and had tried to stay at their current organisation, but the economics just no longer made sense. What a lost opportunity for both sides.

In a similar vein, what if the value of someone's skills, knowledge and experience suddenly rise significantly due to the demands of the external marketplace? 15 years ago when I started working in lean and agile, not many people knew much about them. Over the following 15 years, they have both become hugely more popular due to companies realising their importance, so a factor of my personal current market value has actually risen entirely independent of anything I personally have done. Again, employees who realise themselves to be in these sorts of situations try to make the economics work through negotiation with their employers, but often things like the 10% increase rule make it impossible, and the employee leaves to go and claim more of the value they create by working elsewhere.

So, what do you do about this? It's likely you'll find the 10% increase rule hard to shift initially, as it's hugely ingrained in many organisations. Even examples of the sorts of damage it can cause to employees and organisations, as cited above, are often met with a shrug that they're just 'one of those things that happen sometimes'. However, at the very least you should be aware of this issue, and be prepared to understand where a candidate or employee is coming from in this whole value delivery conversation. In fact, if you were hiring for a lean or agile role, and the candidate *wasn't* having these sorts of conversations when presented with your '10% increase' rule, I'd be slightly skeptical of the degree to which they were truly living the lean and agile values. So be aware of these issues, be open to discussing them and maybe, just maybe, start to build a salary and benefits model that better aligns with the complex, uncertain and non-linear nature of value delivery in the lean and agile space.

Section 7: Growth And Development

Idea 29. Is Your Capabilities Framework Harming Your Lean-Agile Transformation?

Hypothesis: Your current capabilities framework may be set up to reward non-lean and agile behaviours, harming your transformation.

Experiment: Run a desk-based evaluation of your current competency framework with the help of experienced practitioners, comparing it against lean and agile mindsets and behaviours, then start to introduce and measure changes to your current framework.

If you're an organisation of any significant size, chances are that you've got a capabilities framework. A document or policy that sets out the skills and competencies you expect from the people who work for your organisation. Perhaps more specifically the skills and competencies required for people working in each role, and / or working at different levels of the organisational hierarchy. It's likely to be the sort of thing you use to assess people against when filling in their annual performance appraisal, or when deciding who to promote to the next grade up.

I'll be honest. The very latest thinking is that this whole capabilities framework concept is wrong right at its heart. But to overcome those flaws would likely lead to such colossal structural and cultural changes in your organisation that they're almost not worth talking about at this stage. So for now, let's assume that the reality is that you're going to have a capability framework, that you're going to use it for assessing people's performance, and likely their fitness for promotion too. How do you at least make it more fit for purpose for your organisation's transition to lean and agile?

Well, there are two main elements to this.

The first is quite simple. Chances are your capability framework is heavily influenced by your old ways of working. For example, do you reward and promote people…

- who show proactive leadership?
- who demonstrate great influence amongst their peers?
- who step up and take ownership of things to 'drive them forward'?
- who always get their deliveries over the line on time against all the odds?

Well, then you're promoting exactly the wrong sorts of behaviours for lean and agile to succeed. For example, showing 'proactive leadership' is very often the problem within systems thinking that Peter Senge calls 'The Illusion of Taking Charge'. As he puts it;

> *"…proactiveness is reactiveness in disguise. Whether in business or politics, if we simply become more aggressive fighting the 'enemy out there', we are reacting - regardless of what we call it. True proactiveness comes from seeing how we contribute to our own problems."*
> (Senge, P.E., 2006, p21).

Showing great influence amongst peers could be good, but is it genuine influence, or just influence through the exertion of power and assumed authority, which reduces the autonomy, mastery and purpose of those same peers? In short, is 'influence' just a nicer word for the 'command and control' behaviour found in traditional project management?

Stepping up, taking ownership, driving things forward and getting them over the line against the odds ignores the fact that if lean and agile were working correctly, things *shouldn't* be this difficult to deliver, and if they are, that's a problem to fix, not heroism to be rewarded. Promote these 'heroes', and they'll promote similar heroes after them, making it less and less likely that your problems will ever be solved. As Dantar

Oosterwal puts it in his book The Lean Machine;

> *"Most organisations unintentionally create systems that require*
> *extraordinary people to deliver ordinary results."*
> (Oosterwal, D., 2010, p70)

Promote these sorts of people, and you're saying that it's the individual that matters, not the holistic system, and promoting people who will be toxic to such essential ideas as decentralised decision making and individual autonomy. Instead, you need to be looking for and promoting people;

- who serve others
- who have a sense of humility
- who coach and support others
- who let others take the credit
- who do whatever teams need them to do in order to help them deliver
- who optimise the whole system rather than themselves or their teams
- who focus on quality, even if that means missing a delivery date, because they know that quality is the key to speed
- who tell people to go home to their families rather than working late

All of these sorts of things are behaviours found in agile frameworks like Scrum and XP, and they're the new behaviours you need to be promoting. What the specifics will look like in your organisation will of course be up to you, but it's something you probably need to think through and start to make explicit sooner rather than later. After all, people will have had years, even decades, of previous experience as to what to do to get promoted, and that won't shift overnight.

So where does this leave us?

Well, you need to be looking at your current capability framework and asking at the most simple levels whether it rewards and promotes lean and agile or instead more traditional behaviours. To do this, you need to bring in people who truly understand what lean and agile behaviours look like. People who understand systems thinking, and who have no vested interest in maintaining the current status quo around an organisation's capability framework.

You need to work with them closely to fill in the gaps they may have around some of the organisational realities that may initially be difficult to fix. Once you start to identify some of the anti-patterns in your current capabilities framework, start to replace them with ones that promote lean and agile behaviour. What this may look like will differ from organisation to organisation, and the speed with which you can make the changes will need consideration too. But as with all of these things, recognise the problem, understand it, and start to make continuous incremental changes based on feedback.

Idea 30. Why Don't They Just Go And Deliver Something?

> **Hypothesis**: The career of change agents in an organisation often follows the same failure arc, little of which is in their control, but which costs both sides dearly.
> **Experiment**: Hire change agents as a specific and protected role, provide them with clear vision and strong managerial support, and hold a formal review any time any of these parameters change.

Typically when an organisation wants to shift to lean and agile practices, it brings in people to help them do so. Often heralded with a fanfare as the people who are going to transform the department or organisation, there is actually a pretty common pattern I've seen happen to these people time and again, that plays out over the space of about two years, and results in the organisation getting far less value from them than they hoped. I set it out here in the hope that you can learn from and avoid the mistakes of others.

Stage 1 is when the person is hired. They're brought in by people who don't know a lot about what they're bringing in, they just know that they need to bring it in, and they hope that the person they're bringing in is the right one. This sort of situation is perhaps inevitable, but it does a disservice to both sides. For often the change agent has the right skills in some regards, for example good knowledge of what the change in theory should look like, but they lack the cultural fit with the organisation, or the knowledge of the sector the organisation operates within, or any other similar categories that are important for creating a successful engagement within the organisation.

So underneath the excitement on both sides at the new appointment, the seeds of failure have often already been sown. The way to avoid this is to create a much more rounded hiring approach, one that meaningfully tests against all of the likely success criteria for the role, even if this means bringing in external expertise to assist with the hiring, as it so often does when organisations are hiring for change agents. After all, if the organisation understood what a successful change agent looked like, they probably wouldn't need to be undertaking the change in the first place.

Stage 2 is their arrival. As I mentioned above, their arrival is often heralded extensively, building up a sense of excitement and hype at this great new person who has arrived. Lovely as this can be for the egos of those involved, it is actually a fantastically bad idea, as successful change agency often takes time, yet everyone is now expecting great things almost instantly due to the sense of hype that has been built up. Thanks to this stage, it becomes very hard to do what successful incremental change agents often like to do. That is, under promise and over deliver in order to build confidence and win influence. If were introducing a lean-agile change agent into an organisation these days, I'd almost be tempted not to mention their arrival to too many people at all at the start.

Stage 3 is almost the BAU part of their role, the time when they might score some victories, but they will as likely run into some pretty big blockers around organisational structure and inertia. This is where they have to hope that they have the right sort of line manager looking after them. Ideally one that gives them a real sense of vision for what they would like to see achieved, alongside plenty of freedom and autonomy to achieve it how they see fit, but who also has their back as soon as they go to them for help. Often however, these change agents get placed with line managers who don't know how to handle the sorts of skills and behaviours that change agents present, so either they end up being micromanaged, or alternatively just left floating aimlessly around without any meaningful support. In both of these instances you often find the catalyst for the fourth and final stage. For when the line manager doesn't know how to handle the change agent, they may ask someone else to take over as line manager, who may then experience the same problems, and in turn ask yet someone else again to take over as line manager instead.

So then we find ourselves at stage 4. The change agent, having arrived in a blaze of glory, has largely only disappointed people, as they were never able to live up to the myriad different high expectations everyone heaped upon them. They've become somewhat demoralised due to the organisational inertia and lack of support, and they now have a line manager who likely has no idea why they were hired or why they're still in a role that doesn't seem to have any focus. Besides, they've been there a couple of years, surely the change has been completed by now anyway? It is at this point the change agent often hears the death sentence for the role pronounced, as their new-new-new-line manager says to them;

"Why don't you just go and deliver something?"

After all, delivery is what the senior people want to see, and it feels like too much of a risk to the line manager to have someone on their team who isn't delivering anything, so the change agent gets told to go and join a team and get on with delivering something to prove their worth. Only this isn't what the change agent was hired to do. It may even not be something they're any good at. So either they battle on valiantly until they get removed from their role, or they take the hint and start looking for a job elsewhere.

The tragedy is that change agents and people who help others deliver are very often hugely valuable where they are able to do their role properly. However, the common patterns of organisational behaviour often destroy their ability to do their role properly time and time again. So if you find lean-agile change agents in the population mix at your organisation, be very aware of how they're treated through the lifecycle of their employment. Bring them in carefully, using people who understand what they're looking for. Don't set up too high expectations for them that they can only ever fail to live up to, and make sure they've always got the right sort of line management for the role they're carrying out. Above all else though, don't turn around and make them go and deliver something when you've forgotten why they were hired in the first place. If you want someone to do delivery, hire someone who wants to do that role, and leave the change agents to be good at what they're good at too.

Idea 31. Do You Need An Informal Internal Job Market?

Hypothesis: Freeing up people to move to where they are needed in an organisation, without bureaucracy and paperwork, will help foster collaboration, reduce silos and optimise the whole.
Experiment: Start facilitating the free movement of people within an organisation based on needs and skills, and track the degree to which this happens.

Many years ago, I was working in probably the most agile place I've ever worked. We were a small startup, growing well but still small enough to fit the entire company in one medium sized room. At some point whilst working there I'd made the mistake of answering a cold call from a telephone salesman, and as a result had obviously gone down on some list somewhere as being someone worth calling. From then on, the cold callers asking for me came in pretty regularly, but one call really made me laugh.

As the salesman was doing their usual thing of explaining the benefits of what they was selling, he got onto the money side of things, and concluded by saying

"That'll please your accounts department!"

I started laughing, and when the call had ended, I went and shared the joke with Hannah. Hannah was our awesome office manager, Scrum Master and occasional writer of code. She also cared more than most of us that our clients paid us on time, so she was the closest we got to having an accounts department. The idea that she was now actually an entire department, rather than just a Hannah, amused us both.

This incident has always stayed with me, but I'm beginning to realise there was actually something more profound in it. In bigger organisations, we get really used to the idea of there being different teams, different departments and different areas of the business. People work to become managers and leaders of departments, they fight over resources and territory with other departments, and generally ignore the idea that collaboration should work within and across an entire organisation, not just within an agile team.

We're so unquestioningly used to this, that it spills over into our recruiting practice too. When we advertise for a role, we advertise for a specific role, located within a specific team, within a specific department. Right from the start, we're implicitly saying to the people we hire that this is a siloed and territorial organisation, where boundaries matter and dependencies are merely managed, not designed out through optimising the whole system.

This can also quickly lead to bloating head counts and spiraling wage bills too. If your worth as a manager or leader is judged by the number of people working for you, then you're going to want to increase that as much as possible, and you'll be loathe to give up that headcount to other people. I've seen this get so bizarre that one organisation I came across had three people on Linkedin each calling themselves 'Head of Digital Communications' for the organisation. They weren't lying either, three different areas of one single department had hired three different Heads of Digital Communications, each considering themselves the only 'real' one for the organisation.

All of this of course is a huge risk to agility, especially agility at scale. If I think back to when Hannah was our accounts department, part of our agility came from the fact that we didn't really identify as different areas of the business. Sure, there were people who identified more strongly with some types of our work than others, just as Hannah cared more about our invoices getting paid than the rest of us did, but ultimately we all saw ourselves as just working for the organisation as a whole. If someone needed some skills pulled into their area for a while, they didn't go out and hire more people and create their own mini empires, they just asked for help and whoever was interested and available wandered over to help. After all, as a startup, we were at risk of not earning enough to make payroll each month, so silos, large teams and hierarchy were money draining risks to be avoided, not career goals to work towards.

So how can people managers help deal with this issue? It's probably too much to ask to expect them to abolish organisational hierarchies, departmental structures and silos altogether. That sort of heretical thinking would take a large organisation years to implement, if it would ever happen at all.

There are two things people managers could do to help deal with this issue though. The first is setting expectations differently when hiring. Rather than making the job ad and the interview process all about the specific part of the organisation that the person will be working in, instead make it much more about the wider organisation they will be joining. Sure, they will start off by working in one part of it, but you could help set the expectations right from the start that everyone in the organisation is focused on delivering organisational goals, not fighting for or even within a specific departmental silo.

Second, they could make it much easier for people to move around different teams as and when needed. Just as one of us in our startup would freely wander over to help for a bit when someone said they needed a hand. Of course, this happens in organisations at the moment, but it's generally done through a pretty formal and bureaucratic secondment process, where a secondment opportunity is identified, an internal advertisement issued, applications received, interviews held, and a person seconded for a fixed period of time. Hardly the flexibility that allows people to wander over to another part of the business and help out for as long or as short as is needed.

Instead of this, how about maintaining an internal live skills board, where people can register the skills they actually have (not the formal role they hold) so that other people can search for people with the skills they need and approach them? Alternatively, run it as a simple and informal job board, where people can post the skills they need, how long they're likely to need them for, and then allow people to move around between teams freely. No formal secondment period, no applications, no interviews. Just join the new team, see if things work out, and keep things moving.

If you're the kind of organisation that runs on data, or even where people only do things when given targets and metrics, then why not even run a metric around this issue? Track how easily and how often people move around the organisation, and perhaps, carefully, incentivise this sort of behaviour through the publication of data or setting of targets.

So there we are then. An approach to breaking down silos and boosting collaboration and agility, that likely runs at low cost with high benefits for employee satisfaction, flow and value delivery.

Idea 32. Develop A Workforce Fluidity Metric

> **Hypothesis**: People learn, grow and develop by being exposed to new experiences and opportunities, and agile people welcome change. The fluidity of your workforce within your organisation may be a good proxy measure for the organisation's lean-agile adoption
>
> **Experiment**: Start tracking the frequency and degree of movement of people within your organisation, and potentially facilitate and incentivise its increase.

I travel a lot with my work to deliver training courses, and at the end of each day, after my co-trainer and I have debriefed the day and planned the next day over dinner, I sometimes find myself in some random hotel room with little energy left but my mind still racing and unable to sleep. So whilst I wait for the sandman to arrive, I wander around the Internet.

One time a few months ago, I found myself wandering around Linkedin, and thought I'd check in to see what a bunch of people I used to work with were doing now. The job was at a large corporate organisation, and whilst the role had the usual challenges you get in such places, the people I was working with were all really lovely. As I searched through and read their Linkedin profiles though, something struck me. Despite me having left the organisation a few years beforehand, out of the couple of dozen people I checked, only one of them now had a different job title from when I left.

I must confess, I was shocked. These were all great people, and I'd have thought a good number of them would have been promoted by now. Even if they hadn't been, at least perhaps they would have moved into a different role in order to broaden their skills and experience? In the few years I'd been gone, I'd already changed job title three or four times due to wanting to broaden out my skill set, and the promotions you get from taking such an approach.

I got to thinking about it, and realised that this may very well be the difference between agile people and agile organisations, and those people and organisations that are still very much taking traditional approaches. Agile people welcome change. It's one of the ways you can tell that someone has truly adopted the agile mindset, when they respond to change as something exciting and an opportunity for growth, rather than something to be avoided and feared. Equally, agile organisations are typically much happier to let people move around between roles and job titles. After all, as new information emerges, it may be that new roles are needed, or new people are needed in those roles, so as a result, the organisation is happy for people to get up and move.

I started to wonder then if these were phenomena that could be tracked as metrics to assess the success or otherwise of an organisation's agile adoption. Could you track how often people's job titles within an organisation were changing, or the regularity with which job titles changed as a function of people's hierarchical layer within the organisation? Essentially using workforce internal fluidity as a metric for organisational agility.

It would be an easy metric to track, as people's job titles are all typically recorded in a central system. For the same reason, it would also be a difficult metric to game and manipulate. In addition, rather than just being a metric that provides an indicator for organisational agility, it might be one that would help to increase it too. People always focus on the metrics that the organisation is tracking. So if you create organisational job role fluidity as a metric, you will at the same time incentivise other employees and managers to increase their fluidity, creating more opportunities for people to move into new roles to grow their knowledge and experience, helping to break down organisational silos and increasing the degree to which people become cross-skilled and cross-functional. Taken to its greatest extreme, at 100% fluidity, you might even end up in a truly holocratic organisation, where people are free to self-organise into whatever roles they judge the organisation needs at that time.

Of course, to make this happen might require a large amount of work, but it's work that should, to my mind, largely be driven out of the HR department. First of all, is it even possible to track the current state of workforce fluidity, and see how it has or hasn't changed over the last couple of years? If it is, then how can HR help people become comfortable with it as an idea, both workers and managers? Once they're happy that it's the right metric to try to shift, how many organisational impediments and blockers are there in the way to people becoming more fluid in their roles? To start with, I'd wager there are many. After all, chances are that for a long time, your organisation has prioritised order, stability and a very clearly structured approach to role progression. HR will need to get all of these cleared before the promise can start to turn into a reality.

There's also the circle to be squared around fluidity turning into chaos. If anyone could do any role they wanted whenever they wanted, then chances are that the overall system would start to fall apart. How do you give people the maximum freedom and opportunity they need to take an autonomous approach to deciding which roles the organisation needs them to do, and which of those roles overlap with roles they'd like to do for their own personal growth and development? How do you make sure someone leaving a role doesn't leave an important gap that isn't backfilled? How do you deal with the potential loss of benefits you get from no longer having long-lived stable teams, that have gone through forming and storming and up into norming and performing?

I strongly believe there are good answers to all of these questions, but equally, achieving organisational fluidity isn't likely to be easy. However, the rewards for those that do achieve it are likely to be great, and the effort to do so will therefore be worthwhile. The first step though would be just to gather and track the data. See what's going on in your organisation at the moment, and decide from there your own route to creating a more fluid approach, one that truly empowers employees to grow, develop, and do what is right for the company and its systems, based on the unique knowledge they have in their own particular areas.

Idea 33. Use Lean People Management To Reduce Organisational Waste

> **Hypothesis**: Employees often have skills and experience that are underutilised, which is a form of waste.
>
> **Experiment**: Run an audit of skills and experience within an organisation, empowering employees to state how they feel they could best be utilised.

In this idea, I'm looking less at the world of agile and more at the world of lean. The fields of agile and lean production have come closer and close together over the last decade or so, and there's a lot to be gained from that. For example, teams following the Scrum framework are increasingly using lean tools like kanban boards and work in process limits to improve the speed and quality of their delivery.

One of the most interesting elements of lean production though is the idea of waste. Waste is essentially something the customer won't pay for, and it breaks into two forms, necessary waste and unnecessary waste. Necessary waste is something the customer won't pay for but that still needs to happen in order to create the things the customer will pay for. A hiring process would be a good example of this. The customer doesn't get any value directly from paying for your hiring process to be run, but if you didn't run it, you wouldn't have any staff and nothing would be created for the customer to pay for.

However, unnecessary waste is something the customer won't pay for and that doesn't enable the things that the customer will pay for. I'm sure we can all think of examples of this in any organisation we want to look at. Lean production categorises these wastes into more specific forms to help understand them too. Things like work sitting idle before being delivered, defects appearing in the work, people sitting waiting for work to come to them, the list goes on.

There is though one specific type of waste here that is hugely relevant to people management professionals. This is the waste of unused or underutilised skills and knowledge in the organisation. Unless you live in a perfect world, where every person in the company has found exactly the right role for them, one that perfectly utilises all of their skills and knowledge every day, then the chances are that you're currently experiencing a great deal of this kind of waste. Indeed, if you look at what some lean people talk about in relation to wasted skills, it also includes things like employees' previous experiences and their creative ability. In short, there are a huge amount of different skills that might be going to waste in an organisation's talent pool.

So what can people managers do about this? Well, pretty much everything I would say. If there's an area of agile and lean that sits firmly in a people manager's court to deal with, it's this one. When was the last time you conducted a knowledge, skills and experience audit in your team, department or organisation to find out just what might be going to waste? Why not run one, and build a true picture of how your workforce's skill match the skills gaps you think you have? You may be surprised by what you find.

Once you've done that, there will doubtless be lots of work to do to find ways to match these skills, knowledge and experiences with opportunities for their use within an organisation, all without creating huge amounts of organisational disruption, which might in itself create forms of waste. For example, how can people more easily find new areas to apply their skills to? How can they be empowered to share their skills and knowledge with others? More broadly, how can you help your organisation to unleash the creative potential of its employees, empowering people to be innovative and creative using their skills to their best, rather than just being worker resources who plough through the work that has been assigned to them?

There's a big opportunity here for people managers to play a hugely important role at the heart of an agile transformation, with almost limitless potential for creating organisational improvement, and ensuring that the skills, experiences, knowledge and creativity you already have in your organisation isn't going to waste. After all, why would your customers pay you to waste that? They'd much rather pay you to gain the benefits from it.

Idea 34. Learn How To Run Training In The Face Of Uncertainty

> **Hypothesis**: Training is often rolled out using traditional project management approaches, pretending that you can know a year in advance how many people will need training in which locations and by when.
>
> **Experiment**: Roll out training with shorter feedback loops. Plan just three months of training, see what happens, and use the information to plan what training is rolled out over the following three months, then repeat.

On one lean-agile transformation I was running, we'd initially run the whole thing on a shoestring. That is to say, we'd had no budget, no team and no leadership support. Every time we wanted to do something, we had to justify why we should be allowed to do it, in order to receive approval and a small amount of funding, and so on. This was actually a great discipline for us, as it meant we had to live the lean-agile values we were promoting to others in a very real sense. We continuously had to inspect and adapt what we did and how we did it, to ensure that we maximised the value we delivered, and do what we did in the leanest and most waste free way possible.

Just because it was a great discipline though, and one that I believe led to our ultimate success, it didn't mean it was easy. For example, we organised dozens and dozens of training courses and trained over 1800 people with zero support from the organisation's official training department. So we had to do all the room booking, attendee booking and everything else ourselves. Early on I'd contacted the official training department to ask for support, and their first question was to ask whether this was training for business people or IT people. I said it was both, as the whole point of what we were doing was to bring these two divisions closer together, using systems thinking to optimise the whole. They just didn't know how to handle requests like that, as business training was handled by one team and IT training handled by a different team, so I just left it there. Looking back, this should have been my first hint of what was to come.

Fast-forward a couple of years, and our transformation was making real headway. So much so, that we were now being approached by people from all over the wider organisation with requests for help. Some of these requests were also ending up in the inboxes of the formal training department, so they got in touch with us to discuss how they could help to formalise and scale up what we were doing. Fantastic, I thought! Finally, some help with the administrative burden of what we were doing, and hopefully some new things we as trainers could learn from working with a full time professional training department too.

I'll never forget the first meeting we had, where we all got together to discuss how to move things forward. After the initial introductions, they got down to the main item on the agenda. How many people did I want to be trained in the next 12 months, who were they, and where were they based? As an agile person of many years standing, I must confess, I was completely lost for words.

I replied that I had no idea how many people needed training, nor who they were or where they were. We had a rough roadmap for where we thought the transformation may go next, but we were continuously inspecting and adapting what we were doing based on feedback and things we learned. Sometimes great new opportunities emerged out of the blue that we had to follow, whilst other opportunities that looked promising kept falling through at the last minute, because talk is cheaper than action.

In return, they looked completely lost at my answer. All of their training experience to date had been based on knowing how many people needed training in what things by which date and in which locations. As we effectively sat and marveled at the stupidity we perceived in each other, for my side I suddenly realised why so many of the large corporate training initiatives I'd been subjected to over my career had failed. The ones where everyone was suddenly mandated, on threat of disciplinary action for non-compliance, to click through some irrelevant online learning[10], then complete a multiple choice test at the end, for which the answers had already been handed to you by the one team member who'd agreed to sit and work them out. Answer 1 is option A, answer 2 is option D, answer 3 is option B, etc. Or even better, where there was no test at the end, but you just had to click a button on completion of the 'training' to attest legally that you had understood what you had been 'taught'.

I'm sure all of these came from a situation where a senior person needed to prove the organisation was being proactive about rectifying a previous serious issue, and had therefore mandated that X people in Y locations must be trained in how to not repeat this issue by Z date. The fact that the people trained learned literally nothing didn't seem to matter. What mattered was first of all planning the training, then proving that the plan had been delivered.

Our problem was that we weren't looking to tick boxes. We were looking to transform how people acted and felt at work on a permanent basis. That's not work that's predictable up front, it emerges over time based on trying things out and seeing what results you get.

How do you deal with this issue though? It's actually one that has popped up in multiple forms over my career, like the time that I needed an organisation to buy training licenses, for which there was a huge discount on offer for bulk buying. But again, I couldn't *prove* that we needed to make a bulk purchase, I only knew we would need that many through personal experience, so the organisation turned down the discount and slowly bought licenses in small batches, until it had bought as many as were offered with the discount, only with none of the cost

---

[10] I've worked in financial services for quite a lot of my career, but I've never worked in a physical branch of a bank, or even come close to doing so. Despite that, I've been through literally days of mandatory online training teaching me how to spot potential armed robbers before they draw their weapons.

savings.

I admit that my stance on the issue, of just saying that you cannot know up front the number of people to be trained, their locations and the date by which the training must be completed may not seem helpful. Especially not to people who are used to operating from the exact opposite position. However, I think the ability to run training in the face of this complexity and uncertainty is essential to the success of a lean-agile transformation, and there are a number of options that might help bridge the gap between the two worlds.

The first is my preferred one, based on trust. The organisation decides that it wants to be serious about investing in the transformation, and in the growth and development of its people, so it allocates enough budget and capacity to enable everyone in the organisation to do whatever training they decide is appropriate for their role. In the same way the electricity grid in a city is actually only capable of providing a small proportion of the power that the whole city would theoretically need if everyone turned everything on at once (because that never happens), I'm sure over time the organisation would find that it needed far less budget and capacity than the theoretical maximum, as not everyone would want training in everything, but this is only something that could be worked out over time, and would require an initial leap of faith.

A second approach is to base it on training capacity. Rather than work out how many people need training where and by when, instead work out how many people your trainers could train within X amount of time without burning themselves out or dropping their quality, then use that to allocated budget and capacity. After all, there's no point offering more money to something than could physically be spent.

A third approach is to take a lean-agile stance and run small batches with regular reviews. That is, forecast one to three months of training, deliver that, then inspect how it went and use the learnings to plan the next one to three months. It won't suit training functions that believe in annual plans, but it will allow you to deal much better with complexity and uncertainty.

There may be other approaches to dealing with this issue too, and if you've encountered any, do please get in touch to share what you've learned. For now though, it's probably enough that you're aware of this problem and start to think about how to tackle it in your own context. After all, you can't run a lean-agile transformation using big up front project plans, and your training function should have to adapt to this reality too, no matter how alien to them it may initially seem.

Idea 35. Use Cross-Training To Cut Queues

> **Hypothesis**: In a world of limited resources, where not everyone can be trained in everything, 'cross-training' people on either side of a queue formation point may bring the highest return on investment.
>
> **Experiment**: Map out your product development flow, track work through it to spot queues forming, then target cross-training in skills either side of the queue.

As you explore the worlds of lean and agile, it will probably strike you just how much there is to learn. Indeed, you should hopefully realise that the need to learn continuously, to inspect what's going on and make adaptations to your knowledge, skills and practice based on what you discover, are right at the heart of what it means to adopt lean and agile ways of working. However, with all of this comes a problem. Where do you start? Who should learn what and when? How do you decide where to target your training? After all, you likely have a finite training budget, and due to the nature of human existence, you definitely have a finite amount of time.

Well one answer to this question popped out at me when I was reading a marvellous book by Don Reinertsen, called The Principles Of Product Development Flow. I'm not suggesting you add this book to your reading list by the way. It's a phenomenal book that makes you see the world in a whole new way, but, perhaps influenced by the world of lean, not a single word in it is wasted, and consequently it is incredibly complex, dense and challenging to comprehend.

In essence, the book looks at some rules and patterns that help product development to flow through an organisation, from the initial idea right through to the end output being used by customers. As a result it helps to start to think about how that flow might look in your organisation.

What stages do products go through in their development? Who gets involved in what at which stage? What does the end-to-end delivery journey for a product look like when it is being developed? For example, you might have a product development flow that looks like the below;

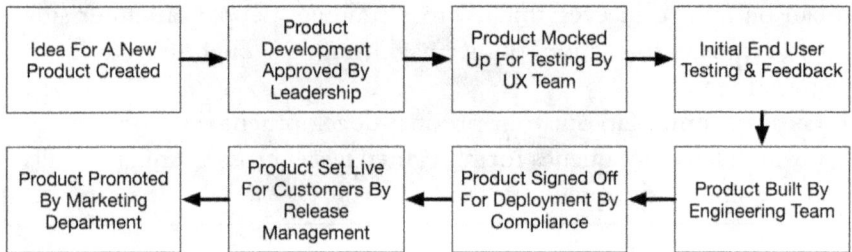

```
┌──────────────┐    ┌──────────────┐    ┌──────────────┐    ┌──────────────┐
│              │    │   Product    │    │ Product Mocked│   │              │
│ Idea For A New│──▶│ Development  │──▶│ Up For Testing│──▶│Initial End User│
│Product Created│   │ Approved By  │    │ By UX Team   │    │Testing & Feedback│
│              │    │  Leadership  │    │              │    │              │
└──────────────┘    └──────────────┘    └──────────────┘    └──────────────┘
                                                                    │
                                                                    ▼
┌──────────────┐    ┌──────────────┐    ┌──────────────┐    ┌──────────────┐
│Product Promoted│  │Product Set Live│  │Product Signed Off│ │              │
│ By Marketing │◀──│For Customers By│◀─│For Deployment By│◀─│Product Built By│
│  Department  │    │   Release    │    │  Compliance  │    │Engineering Team│
│              │    │ Management   │    │              │    │              │
└──────────────┘    └──────────────┘    └──────────────┘    └──────────────┘
```

Once you've got this understood for your organisation, or at least understood one of the product development flows that exists in your organisation, as there might be a number of different ones running at the same time, you can start to measure how work flows through it.

For example, you could take an idea right at the start of the flow, and see how it moves through the different stages. Does it take longer to pass through some stages than others? Does it sometimes move backwards rather than forwards through the flow when mistakes are found to have been introduced at an earlier stage and need to go back to be corrected? Do ideas get to a certain stage then get abandoned? The idea is to look at this whole system and optimise the flow of work through it, so it is as smooth, inexpensive, simple and speedy as possible. Once you've got this working, innovating by getting new ideas into the hands of customers can occur at a much faster rate for much lower cost, giving you a significant competitive advantage. This is the benefit of mapping and measuring flow.

But what has this got to do with training? Well, Don Reinertsen is strongly interested in queues of work. Queues are typically formed of work that is sitting waiting to be processed at a certain stage in the flow, and generally we try to avoid them. Work sitting waiting for something to happen to it is a form of waste. It has cost money to get into the queue, it's costing you yet more money by sitting in the queue, and perhaps most important, it's not making any money whilst it's sitting there either. Now these queues can appear for a number of reasons, but one common reason is that the stage of the flow where the work is piling up into a queue is one that needs specialists to work on it. For example, it may be a stage where a certain sort of testing needs to be carried out before the work can move onto the next stage. You may only have a small number of these testers, and they only have so many hours in the day, so work begins to build up for them in a queue, waiting to be tested.

One solution to this then might be to hire more testers. More testers will lead to more testing capacity and reduced queues, right? Well, possibly not. You see, as we looked at in idea 6 this book, adding more people to a piece of work can actually increase its complexity and so slow things down. Not to mention the fact that the new testers are going to take some time to get up to speed with things, so even if they don't slow things down, you're not going to see an immediate benefit from hiring them. Besides, hiring more people costs more money, and the queues at that stage in the flow might only build up occasionally, making hiring more people not worth it economically. So what do you do then?

Well, this is where Don Reinertsen suggests 'cross-training' people. For example, if your flow diagram shows a stage before testing called development, and a stage after it called regulatory sign off, then why not target the developers and the regulatory sign off people with training in how to carry out testing work? They're closest to this stage in the overall flow, sitting either side of it as they do, and so they should have some awareness of the work that goes through it anyway. They've also got the most 'skin in the game' when it comes to that stage, as when the testers are maxed out, the developers can't do more development without building up an even bigger queue, and the sign off people are sitting there without enough to keep them busy, due to the blockage in testing.

This is a reason why an agile framework like Scrum doesn't have different job titles in it beyond 'Development Team Member'. The team may contain skills for various different stages of the overall flow, but increasingly, team members are expected to pick up skills and knowledge from others in the team, so if there's a queue building up at one stage, they can pitch in and help clear it to keep the work flowing.

Now achieving this won't be possible for people managers alone. It is almost certain that other people in the organisation will need to be involved in mapping out what the end to end flow looks like. However, people managers should be close to this work, as they can learn a lot from it when it comes to deciding where to target training efforts. Look for the queues, and look to cross-train the skills that are causing the queue with the people either side of it. That way you don't have to hire additional people and suffer the increased cost and temporary slow down in work that they cause. You can also promote the idea that everyone is part of a wider system, reducing siloisation of thinking and sub-optimisation of the whole by raising everyone's eyes to see what other people within the system do. It'll take time to achieve, and require collaboration with other parts of the organisation to get started with, but the benefits could be significant and long lasting.

Idea 36. Understand Who Is Really Training Your People

> **Hypothesis**: Online training platforms have an incentive to produce high volumes of material at low cost, leading to lowered quality standards that harm the training your employees are being offered.
> **Experiment**: Consider the cost / benefit analysis of saving money on outsourcing employee training to online learning platforms, and at the very least, review and audit any content on the platforms before it is offered to employees.

I have a theory. The larger an organisation is, the more structured it tries to be, and yet the more chaotic it actually is. Now I don't mean chaotic in a negative sense, as very often you need some degree of freedom, brought about by chaos, to allow organisations to inspect, adapt and improve how they operate.

However, the danger for me comes when organisations *think* that they've got a handle on something through policies and procedures, and as a result, they take their eye off the ball, allowing problems to emerge and fester away unnoticed. I suspect that the problem I'm about to discuss comes from exactly this sort of source.

When it comes to organisational processes and performance, it's only natural that organisations, especially large ones, will want to maintain a close degree of control over how they operate. Part of how they operate is how the knowledge of their operation is spread and maintained across all of the relevant employees. As a result, organisations rightly keep a close eye on how people are being trained. An unfortunate side effect of this phenomenon is that sometimes organisations keep such tight rein on it that they in effect shut down all but the most mandatory training. After all, if people aren't being taught anything, then they're not being taught anything wrong are they? In addition, the mandatory training they do get becomes so controlled that it becomes entirely standardised and essentially a tick box exercise, where organisations are required to demonstrate that people have attended the training, not that they have actually learned anything as a result of it.

Where organisations do allow a little more freedom around training though, perhaps through bringing in external trainers, or hiring trainers as full time employees, organisations still want to keep a tight rein on what these trainers are teaching people. In one sense this can actually be quite beneficial. For if the organisation is keeping control over what is being taught, those being taught it can have some confidence that what is being taught is mandated to be a reality within the organisation.

For all of this idea that organisations keep a tight control over what is being taught though, there is I suspect a growing problem, caused by the blindness organisations experience through believing that everything is under control.

Increasingly, online platforms are building up large knowledge banks of training courses, and selling them to organisations on various different licensing models. The appeal for the organisation is clear. They can outsource something they don't particularly want to run themselves. They can have people sit and learn at their desks rather than taking time away from 'productive work' (because of course, learning is never seen as a productive thing to do). On top of all this, they can sit back and believe that they have access to a whole host of training provided by industry experts, at a far lower price than bringing those experts in to train their staff directly.

Sadly, most of these views are mistaken. On a simple level, there's the fact that people learn in different ways, and learning by sitting at a desk, by yourself, with head phones on, is likely to be a pretty poor way to learn anything new. True learning comes from a mix of hearing and experiencing, through approaches such as discussion, games, simulations, reading, writing and all the rest. The idea that you can train up a workforce through buying in some online tutorials is entirely missing the bigger picture.

However, there's an even bigger issue here. The companies that are building these online learning platforms have two big incentives driving their activity. The first is to keep their offering as comprehensive as possible, so that customers don't have any incentive to look elsewhere. The second is to keep the profit margins as high as possible. If there's a natural market limit on what customers will pay in this space, then all they can do is to cut costs on creating the material, by hiring people who are less than industry experts. So with an incentive to ship high volume at low cost, the quality of their training material naturally falls.

I assume this is the phenomenon I'm now noticing in various online courses that are being provided around lean and agile. I've reviewed many of them, from many different suppliers, and honestly, the majority have had me squirming in my seat at how wrong they're getting things. There are courses out there teaching people aspects of lean and agile that are just plain wrong. Not wrong in terms of nuance or points of debate. Wrong in the sense of actually factually wrong at a very basic level.

Now, this isn't an argument against bringing in an online content provider to provide training content for your organisation. As part of a well thought through and blended approach to learning there may be significant value in it. However, it is an argument for doing due diligence on what is being provided, and ensuring that you don't have inattentional blindness to what your employees are being taught around lean and agile, just because you think the issue has been resolved. If you don't, then you could have all sorts of confusing, mixed or just plain incorrect messages going out amongst your employees. You spend time and effort vetting the training that's being delivered via your existing approaches. Why should the training delivered via new approaches be treated any differently?

Idea 37. Get Your Leaders To Teach

Hypothesis: Training around organisational change would be far more impactful if delivered by senior leaders, and the senior leaders would receive huge amounts of beneficial learning in return.
Experiment: Find a senior leader who is prepared to run a single one or two day training course, and evaluate the value they gain from it and the impact it has on participants.

When it comes to running training and learning in your organisation, who is it who's delivering the courses and workshops? Members of staff with a bit of free time on their hands now and again? Full time members of staff whose job it is to run nothing but training courses? Perhaps external third party trainers, brought in specifically to run certain courses in their area of expertise?

What if I suggested that actually, one of the best people to deliver your training courses was your CEO?

This may sound pretty strange. After all, your CEO is probably one of the busiest people in your organisation. Surely they won't have time to take a day or two days out of the office to teach some people new skills or ways of working? This may seem like the case, but actually I'd argue there is huge value in them doing so.

First of all, there's the value for them. I've been running training courses for many years now, and I still learn something new with every course I teach. When you want to accelerate your personal growth and development, there's nothing like standing up in front of a room full of people and having your knowledge tested in all sorts of ways you didn't expect. Indeed, I'd recommend that everyone should run some training of some sort as part of their role, no matter where they work, in to help them order to grow and develop.

There is also another unique piece of value more senior leaders could get from running training, and it comes from the idea of gemba. Gemba, in the world of lean thinking, is the name for 'the real place' or 'the place where the work happens', whether that's the factory floor, or the part of the office where the developers sit. Only by going to the gemba can you really see what is going on, really understand the problems that are occurring, and see the truly best solutions for them. Visiting gemba is far more powerful, insightful and useful than reading PowerPoint decks or Excel spreadsheets when it comes to understanding the reality on the ground, and it's the reality on the ground that leaders most need to understand when making decisions.

So many times when I've been running training courses, a delegate or table of delegates has started questioning the material I've been presenting by saying that it sounds great, but that it wouldn't work in their area because of X problem or Y situation. As a trainer, I can help suggest potential solutions to the issues they raise, but if there were a senior leader of an organisation standing in my place, they might be able to take the problem the person is reporting and actually fix it themselves. The more time leaders can spend in gemba the better really, and running a training course for a few dozen people from across an organisation provides an excellent cross-section of gemba to hear from, learn about and do something to help when the leader gets back to the office.

On top of this, another relatively common bit of feedback I get during training courses is someone looking up at me skeptically, often whilst looking at their emails on their phone, and saying;

> *"Yes, yes, this all sounds great in theory, but we're not actually going to be doing any of this in our organisation are we?"*

When training in an external organisation, this is a tough question to answer, as you often have little context about the organization, and even less influence on whether or not it adopts what you're teaching people. When training within an organisation that employs you, this becomes a little easier, as you often do have some context and influence, to reassure people that the new ways of working are real and are supported by senior management. However, if a senior leader was delivering training and was asked that question, all they'd really have to say is;

"Yes, we are."

and that would be that. After all, if the CEO says we're doing this, then it's hard to argue that the senior leadership support isn't there for it. How much more impactful would training around organisational changes be if it were the senior sponsor for the changes were standing up there to explain them to people?

Despite these benefits, this all may still sound like a crazy idea, but in reality, some great leaders over time have got themselves down on the ground to ensure things are done right at the lowest level. For example, Steve Jobs at Apple is said to have taken personal charge of designing the induction program for Apple employees, right down to choosing the venues people were to be trained in. He saw the huge importance of setting the right scene and message for people when the joined the organisation, so he ensured that he had good oversight of the details within the induction itself.

The reality of course is that leaders likely won't be able to do this sort of work full time, or with the sort of capacity that a regular trainer would, or else they'd stop being a leader and start being a trainer instead. However, I don't think it would take many training sessions in new ways of working to be run by senior leaders to send the message out that the organisation was serious about the change, and to give the leader enough access to gemba to start observing some real problems and identifying some real, meaningful changes. They'd need support of course in learning how to teach, and their diaries would need to be rearranged slightly, but I firmly believe there's huge value in the ideas of leaders who teach, for all of the parties concerned.

Idea 38. Teach People How To Share Their Knowledge

> **Hypothesis**: A true learning organisation is one where every employee feels capable and empowered to share their existing knowledge and new things they learn with others around them.
> **Experiment**: Set up a programme to teach people how to run training or knowledge sharing sessions, and make it open access to all.

In working on this book, I have been struck by just how difficult it would be to overstate the importance of people managers in organisational learning for a successful lean-agile transformation. Building an organisation that continually learns, grows and evolves it at the very heart of what it means to be lean and agile.

Now, exactly how to do this isn't going to be covered here. This book gives you ideas to experiment with and learn from, not prescriptions to follow. However, in a simple form, I suspect organisational learning could be broken down into four different levels.

1. Little to no organisational learning takes place, beyond 'lessons learned' documents being written then filed away and ignored at the end of a project. Little to no training is available.
2. Some organisational learning takes place. Training paths are designed up front and prescribed to different roles and job grades.
3. Individual/team-based learning takes place, through a regular process of introspection and retrospection.
4. True organisational learning takes place, through a focus on the organisation as a whole system.

I've definitely experienced organisations that sit in either 1, 2 or 3. Option 1 is the easiest route, and is typically found in what Dr Ron

Westrum called 'pathological organisations', where information is power, so it's hidden away or used for political purposes, alongside a culture of fear and threat. You don't want people to learn too much, as they might challenge the status quo. You certainly don't want to be seen to fail, so any opportunity to learn from failure is removed as well.

Options 2 and 3 are where most organisations start off with their lean-agile adoption, and to be honest, it's where many organisations end up as well. Teams run retrospectives on how they're working every two weeks or so, and they try to do things with what they learn from them, but essentially what they're doing is more about following the new 'agile process' that's been prescribed to them, rather than actually learning, growing and improving. Westrum would have characterised these places as 'bureaucratic organisations', places where defined process are king, and must be followed. Even failure will have a defined process by which it must be judged.

Option 4 though is where you want to end up, an organisation that genuinely learns, grows and evolves in response to changes in its internal and external environment. What Westrum called a 'generative organisation'.

One of the shifts between a generative organisation and the other forms of organisation is in the usage of information and knowledge. In both the pathological and bureaucratic organisations, there are typically restrictions on how and what knowledge can be shared. In generative organisations, knowledge can be shared freely for the benefit of all. Which creates another interesting problem for creating such an organisation. What if your people don't know how to share knowledge?

I suspect a lack of ability to share knowledge is one of the often hidden reasons behind the failure of a lean-agile transformation, and one in which people managers could play an absolutely critical role. There are a number of elements to it.

First, are you building a culture where the learnings that people generate are actively shared by default? Not just having a fixed 'lessons learned' template to fill in at the end of a project, which is then uploaded to an Intranet site, but actively collating learnings that are generated, analysing and synthesising them, then sharing the salient points proactively across the entire organisation. This doesn't have to be a difficult task. In bureaucratic organisations the information is probably already there, it just needs some people to unlock it and bring it to life.

Second, are you enabling people to share their learnings far and wide? Someone collating, analysing and sharing learnings from across the organisation is a good thing, but the learnings would be far more engaging and powerful if they were being shared and taught by the people who discovered them, giving learners the ability to ask questions and engage in debate around the learning too. How many people are naturally gifted public speakers and teachers though? Not many, especially if you've not been looking for those sorts of skills when hiring people. So again, people managers could play a crucial role here in teaching people how to teach, turning everyone in the organisation not just into a learner, but also into a teacher, someone who shares their knowledge as often as possible.

Organisational learning isn't just about teaching people through training courses and prescribed processes. It's about creating an organisation where the boundary between learner and teacher is blurred, and the knowledge in everyone's head is unlocked and shared for the benefit of all.

Idea 39. Make Sure People Learn Things

> **Hypothesis**: Organisational learning is hugely valuable, but is often the first thing to be cut under the pressure of delivery.
>
> **Experiment**: Decentralise decision-making around what to learn to individuals and teams, mandate a capacity allocation policy for it alongside their other work, then make their learning transparent through backlogs and demos.

Organisational learning and development sounds great, doesn't it? The more people learn, the better they will be, and the more value they can deliver to the organisation. Organisations spend millions each year on helping people to learn. The only problem is, at least from my experience of working in large enterprises, very often people don't actually learn anything at all.

So often, the millions invested in learning are spent with large suppliers providing big, complex Intranet based learning systems. Being the 'one size fits all' approach that they generally are, the learning materials that get put on them tend to be very generic too, and only contain the sort of content that the organisation needs to be able to *prove* it has taught to people. Health and safety, non-discrimination, regulatory compliance, that sort of thing. The problem is, being online based learning, it's very easy to just click through this sort of thing and register completion without even reading anything, let alone learning anything. Beyond ticking a box to show some form of due diligence has been done, what's the actual value of approaches like this?

However, in so much of lean and agile, we look to decentralise control back to teams and individuals, so why should learning be any different? This would be the opposite approach to what I've described above, but could potentially bring huge benefits. Individuals and teams have better visibility over what they need to learn in order to succeed in their own specific contexts with their current skill sets, so why not let them decide what to learn? Now of course I can see objections arising if people were just given a cash allowance each year to spend on learning whatever they wanted, because you may get slight edge cases where people learn things not at all relevant to their role. An accountant decides to spent their budget on gardening classes for example, because they need to tidy up their herbaceous borders at home. Not necessarily the sort of thing an organisation should be expected to pay for.

However, do we trust the people we employ or not? In an agile world, we do, and edge cases are sometimes the small downside we pay for the huge gains that increasing trust and empowerment can bring. Besides, if you really were concerned about this, how about offering the courses that the organisation believed to be relevant internally at a discount on the external commercial price, using internal trainers? That way you can financially incentivise the training you think may be relevant, but at the same time trust and empower those you employ to spend the money as they see fit.

There's also another common issue here that needs to be addressed. All too often, under the pressure of delivery, learning and development is the very first thing to be cut. If not in terms of budget, at least in terms of the time people are given to undertake it. I've trained thousands of people in organisations now, and a really common theme is people either pulling out of the course in the days beforehand due to 'pressures of work', or being physically present in the training room but constantly on their phone or laptop, still doing their day job and mentally absent from the course.

When working as a Scrum Master, I was once even banned by a team manager from taking the team out for an hour to run a simulation to help them better understand how to resolve an issue they were experiencing with their delivery. No matter how much I tried to explain that the team were sitting there blocked and unable to work for hours on end by this issue, and that taking the team out of the office for a single hour might fix it, the manager was adamant that learning was a waste of time when the team should be sitting there 'working'. I suspect this experience is all too common, in a world where people still think being present at work matters more than delivering value at work.

So, how do we deal with this? Well, like anything else in a lean or agile world, if we want something to happen, we need to make it visible, track it and evaluate it once it has been completed. In short, learning and development work should be being added as items to a team's backlog just like any other work they're being asked to complete, and should be demonstrated at the end of whatever time period they're using for getting feedback on the things they've done. If people are struggling to find time for it, then perhaps a capacity allocation policy might be useful to implement too, guaranteeing the team a set amount of time to set aside to undertake training within a certain time period.

Of course, valueless mandatory online training might use up that allocation if you're not careful, but then if it really is valueless, and team members didn't get to choose to take it, then those things should be being called out and acted upon when the learning is demoed in public. The capacity allocation could be set at the team level, but ideally it should be set at a higher level, to mandate that everyone in the organisation sees learning and development as just as important a task as any other task they're asked to perform.

A lot of these ideas are admittedly counter to the way training is usually administered within an organisation, but that's likely because training is generally run in a very waterfall traditional project management kind of way. Taking a truly lean-agile approach to it could pay huge dividends for team autonomy, empowerment and self-direction though, giving people the tools and opportunities they need to solve problems themselves.

Idea 40. Understand The Context Of Agile Certifications

> **Hypothesis**: All agile certifications are valuable to some degree. Organisations would benefit from understanding each of them in greater detail.
> **Experiment**: Map out which certificates are relevant for which level of role within your organisation, and make it publicly visible to everyone concerned.

Agile certifications, for those who have not come across them before, are the certificates you are awarded for attending a training course in agile or one of its associated frameworks. There are lots of frameworks out there, and many have multiple different organisations offering certification in them, so the number of certifications out there runs into the hundreds.

Primarily, it seems that there are two schools of thought around them at the moment. On the one hand, you have the experienced agile practitioners who think that certifications are largely a waste of time, or are at best irrelevant in the face of experience. After all, why should you be 'certified' to do something after just one or two days of training? On the other hand, you have people desperate to break into the agile job market and gain as many certifications as they can, who often put their certification letters after their name on Linkedin and on their CV, as if a two-day course is equivalent to a three-year university degree.

I've always thought that both schools of thought are wrong. Of course, getting a certification from a two-day course doesn't mean you can do a job to a superb standard. Experience still counts for a huge amount. Equally, getting a certification often means that you've had to undertake some new learning, and more broadly that you're the sort of person that is looking to develop and improve yourself. That isn't something that should be belittled or ridiculed.

Reflecting on both sides of this argument over the past few years, it has struck me that what's missing is context. An emphasis on context is something other certifications outside of the agile community have, and it's something we might benefit from bringing into the agile community in turn.

Think of the example of a university interview. Perhaps an interview at the University of Oxford, one of the greatest universities in the world, just to push the example to an extreme. A nervous 16 year old goes up to Oxford for the three day interview process to gain a place to study there, and the only qualifications they have are their GCSE exam certificates[11]. Unsurprisingly, the interviews they have at Oxford don't start with the professor saying this;

> *"It says here that all you've got is some GCSEs. That's rubbish! I'm a professor at Oxford University. I've got way more knowledge and experience than you. I've advised governments across the world and done work that has changed the way people see my whole field of research! You haven't. What makes you think having some GSCEs makes you worthy of being here at Oxford with me?"*

The professors interviewing them don't ridicule or belittle the interviewees for the fact that they only have GCSEs, even though the professors are qualified and experienced to hugely higher levels. They appreciate that GCSEs are as much as this person has had a chance to study for so far, and they look at their results to date, along with their critical thinking skills and general academic potential when deciding whether to offer them a place at Oxford.

Equally, from the other perspective, the interview doesn't start like this;

> *"Well, it looks like you've got some pretty good GCSEs here, would you like to take on the lecturing for my 'Contemporary Critiques in Neoliberal*

---

[11] For non-British readers, GCSEs are General Certificates of Secondary Education; national exams taken around the age of 16, the results of which are taken into account by universities when deciding to whom to offer places.

Institutionalism' unit next Michelmas term? I've also got some evidence gathering sessions with a government select committee coming up, do you think you could cover those for me too?'

Yet all too often in the agile space, we're doing one or the other of those approaches. We either belittle people who are trying to better themselves with agile certifications, or we think that having a certification after a two day course makes them eligible to do a job requiring huge levels of knowledge and / or experience in order to be done well.

What we need to do is to provide context around the certifications, so people understand that it's great to work towards them, that there's value in doing them, but just having them doesn't mean they can do the job they're 'certified' to do, and that the certification is often the very beginning of the journey, not the end.

So on the one hand, the people who belittle agile certifications need to stop doing so. Whether that's humble bragging their pride that they know more than someone with a two-day certification, or open pride that they've never gained any certifications because it's all about experience. On the other hand, people gaining certifications need to realise that there is a whole load of learning to do beyond their certificates, and that that will come through a mix of on the job experience (or research as it would be called in a university context) and studying for further certifications, both within and outside of the agile community.

Separate to this argument in the agile community though, people managers and HR departments need to understand what agile certificates are, the work required to gain them, the difficulty in achieving them and their popularity within the community. Once they've got this, they can then gain a clearer view of both the likely skills and experience of candidates applying for jobs, and also the value of bringing certificated training into their organisations as well. I'd love to see organisations develop a clear hierarchy and set of learning pathways for agile and its frameworks, one where we see entry level certifications as holding definite value, but being what they are, entry level certifications, not sufficient knowledge to advise and coach others across a range of different scenarios and contexts.

I've made a start at pulling this information together at www.humansarenotresources.com/certification, so you can see at a glance what certificates are out there and to some degree what they mean. Feel free to use it to start to build your own organisational approach to the certifications you value, so you can use certifications for their true purpose; being useful building blocks for ever greater learning, growth and development.

Idea 41. Create An Internal Certification For Advice

> **Hypothesis**: A lack of quality control around advice, coaching and consulting within your transformation could be holding it back.
> **Experiment**: Design an internal certification process for evaluating and making transparent the knowledge, skills and experience of lean-agile practitioners within your organisation.

One of the biggest headaches I've experienced as someone running lean and agile transformations is the question of who to allow to be involved in the transformation itself. There are three things at play here.

The first is my own internal biases, neuroses and fears. As someone running a transformation I want it to succeed, and like any human being, my attempts to make it successful risk spilling over into being a control freak, pushing people away, taking on too much myself, overloading my work-in-process limits, delivering less and burning out. That's not your problem I know, but I just wanted to call it out here, as it's something to be constantly vigilant for when looking at the people involved in a transformation. Passion and enthusiasm can often, accidentally, lead to control-freakery, burnout and failure.

The second issue is more relevant though. At some point, you want the transformation to be self-sustaining. In fact, it shouldn't even be a transformation any more, it should just be 'Business As Usual' (BAU), the way the world is, as if there were nothing unusual about it at all. For this to happen though, ownership of the new ways of working have to be totally devolved to everyone in the organisation. The transformation team should at the very least have faded into the background, if not disbanded itself altogether. This should be the ultimate goal of any change agent, and it's important to bear that in mind for this third point.

The third point is the most difficult. You want to hand over ownership and control of the new ways of working to other people in the organisation, but what if you're handing them over to people who don't know what they're doing, or who have a vested interest in maintaining the current status quo?

This may sound like paranoia, but it's not. There are thankfully very few roles in society that require absolutely no proof of a person's competence in order for someone to hold that role. For example, a medical doctor has to prove they are qualified in medicine, and be licensed to practice. A lawyer too has to have a license to practice, that can be revoked if they mess up. Even someone driving a car has to pass a test and maintain a license, which again can be revoked if things go wrong.

There are though exceptions to this sort of approach in some career paths. Politics is perhaps the most noticeable one, where you could in theory be in charge of an entire government department or an entire country with absolutely no suitable qualifications or experience at all. The downsides of this sort of situation though are perhaps mitigated by the benefits to democracy that this approach brings. If we want to have a country that is ruled by its people, then we have to be able to let the people rule, no matter who they are.

One field where the downsides of lack of verifiable experience exist but without any upsides to offset them is the field of lean and agile. Due to the proliferation of the certifications industry in agile in particular, just a two-day course and a simple online quiz can give you the ability to say you're qualified to be a scrum master. Becoming an 'Agile Coach' is even easier. You just add the words 'Agile Coach' to your CV. If you think I exaggerating, I'm really not. Every day I marvel at the people on LinkedIn who describe themselves as 'Agile Coaches' and yet regularly display a complete lack of even basic agile knowledge.

This does become a real problem for an organisational lean or agile transformation though. For the change agents at the centre can't do it all, and ideally they want to be training up other people in the organisation to act as change agents too. But what if the people being put forward as change agents just don't get it, and could actively take the transformation backwards?

Well in once sense, as I noted above, the people at the centre do have to be aware of their own biases in this area. I sometimes find myself pontificating about the non-agile actions of other change agents, only to be gently reminded by a good friend and mentor that I used to make exactly the same mistakes when I started out too, and of course he's right.

However, when you really think about it, the risk isn't so much in people getting things wrong, it's in people getting things wrong whilst holding themselves out to be experts. Which brings me to the idea I think may well solve this whole situation.

There are of course different lean and agile certifications that pretty much prove beyond all doubt that the holder knows what they are doing, but understandably these are quite hard to achieve, so as a result, very few people hold them. In case you're wondering, at the time of writing I'm thinking of certifications like Certified Scrum Trainer (CST) and Certified Enterprise Coach (CEC) overseen by the Scrum Alliance, or SAFe Programme Consultant Trainer (SPCT) overseen by Scaled Agile Inc. These certifications take years of experience and hard work to achieve, and are sometimes seen as the equivalent of a university degree in their intensity and challenge. There are then the certifications that take very little to achieve, perhaps just a two day course, sometimes even with no exam at the end, course attendance alone automatically granting the attendee the certificate.

What an organisation might need is something in the middle, and something that takes into account the organisational context at the same time. A way of certifying its change agents internally to demonstrate their level of experience and understanding in the changes the organisation is aiming to make. Most crucially, these certifications must be regularly checked to ensure that the holder still meets the criteria, and they must be able to be revoked if they do not, or if they commit a severe transgression, just like a driving license.

Only with a system that allows people to verify the skills and experience of the change agent they're working with will any sort of coherence be able to be applied to an organisational transformation rollout. Similarly, along with the stick of removing people who don't know what they're doing from claiming that they do, a decent internal certification framework would also give people a clear progression path from novice to expert, with targets to work towards and recognition to receive in return, creating a strong incentive scheme around the transformation too.

Your transformation is important, and it is to some degree unique. Setting up an internal certification scheme to govern its rollout could respect and advance both of these things, whilst helping to ensure that it maintains its quality at all times.

Idea 42. Fix The Failure Of Self-Organisation

Hypothesis: People and teams don't know how to self-organise. Often they've got no experience or confidence in doing so at all.
Experiment: Run training in team working and self-organisation, and have workshops ready to use to step in and mediate potentially difficult self-organisation conversations that may emerge.

It's funny, one of the most common statements about agile ways of working is also one of the hardest to achieve, and the one of the rarest to see actually occurring. It comes from principle 11 of the agile manifesto, which states;

> *"The best architectures, requirements, and designs emerge from self-organizing teams."* [12]

If ever there were a single sentence that has caused immense amounts of confusion and argument, this is one.

In principle of course, it's perfectly correct. If you let people take ownership of how they work and the outputs they create, then their motivation will be higher, the quality of the work they produce will be higher and the overall end result will be better, as you're letting the people who both know the most about the work and are the closest to it take the decisions. Someone coming in from elsewhere and telling people what to do and how to do it will have less complete information, and also likely have less expertise in the work being done, so their suggestions (or commands...) are quite probably sub-optimal.

The problem with the sentence though is that the phrase 'self-organising teams' can mean pretty much anything to anyone. Of

---

[12] http://www.agilemanifesto.org/principles.html

course, teams should have some autonomy. In fact every team already does. When you're using your computer, does someone tell you exactly on which seconds within a day you should click your mouse button? No? In that case you have some autonomy. At the other extreme though, what if a team decided it wanted autonomy over how much it was paid and so just set its own astronomical salary, then substantially cut its working hours before beginning to work on something the organisation didn't need anyway, and to cap it all off decided to sit there completely naked whilst working on it? Perhaps that sounds like too much autonomy?

So the answer then probably lies somewhere in between. Teams should be free to self-organise on some matters, but other matters may have to be decided centrally if the overall system is to operate in a coherent manner.

The other problem though is that people in reality tend to be quite bad at deciding how to self-organise. After all, it's just not how we're brought up. As children, our parents care for us by giving us a routine, teaching us right from wrong, telling us what we can do and what we must not do. At school, this pattern is only accentuated, with school rules, uniforms, discipline and other strictures that constrain what we do and how we do it. The entire exam system, whilst there to validate our learning, has the perhaps unwitting effect of making us feel like if we jump through hoops that other people tell us to jump through, we will be rewarded. So it should not be a surprise that the modern workplace feels very much like this too, a place where we're placed within a clear hierarchy of command and told what to do.

It's no surprise then that telling people they are now 'free to self-organise' often fails. They just don't know how to do it. More than that though, over the years I've often felt like telling people to self-organise is actually a form of emotional violence towards them. I've found it more noticeable in some cultures than in others, but if someone's been brought up for their entire life believing that a boss must tell them what to do, and suddenly someone turns up telling them the boss isn't allowed to do that anymore, they must decide themselves, then by making them make their own choices, they feel faintly but constantly terrified that they're doing something wrong. As if they're just waiting for the boss to come back and punish them for their free thinking. To give a real example of this, I was training and coaching teams in agile and self-organisation abroad, when I found out that the team manager was going round and telling people, speaking in the local language so I couldn't understand them;

"These guys fly home on Friday. Humour them for now, but come Monday, I'm back in charge again, and what I say goes."

So what has any of this got to do with people management? Well, so often, this self-organisation issue is just left unresolved in a lean and agile transformation, meaning the reality on the ground quickly reverts back to command and control, or alternatively the whole situation descends into arguments and chaos, as everyone argues for their own version of self-organisation, whilst claiming that anyone telling them to do something differently is 'not being agile'.

A people manager's role then in this is twofold. First, individuals and teams need to be trained in self-organisation. It's not something we're taught at home or at school, and it's probably not the current cultural climate of the organisation either, so you can't just tell people to be free and expect them to go along with it. If you want people to form self-organising teams, you have to teach them how to do so.

Second, there will probably be some mediation work required. It would be lovely to think that people could just attend a course in how to self-organise, how to make trade offs and build consensus within a diverse team of people, but realistically it's not going to happen. The behaviours have just been ingrained in us too deeply. However, training followed up by a workshop or set of workshops for a team, to work them through what self-organisation would realistically look like in their context, perhaps with a roadmap for increasing and refining it over time, could be hugely valuable. Who better to sit and act as neutral mediators for this work than people managers teams themselves?

The continued problems and failures of self-organisation within the world of agile have been a problem for far too long now. People managers are uniquely placed to get involved and start to resolve them. It would be fantastic to see them do so.

Idea 43. Celebrate Failure

Hypothesis: Cultures that fear failure and blame people for it are unable to inspect, adapt and improve, and at the same time significantly increase their risks.

Experiment: Drive a change to a culture where failure is approached in a no blame manner, and can be openly discussed in order to learn from it.

When I run training courses, I occasionally get funny looks. Sometimes, when we get on to discussing the idea of failure, I happily exclaim *"failure's awesome!"*. At which point, a whole bunch of people in the class look at me like I'm mad. I can understand why.

You see, I'm not sure what this is like around the world, but certainly western and western influenced education systems bring people up to avoid failure. Right from an early age we're taught to seek our parent's approval, to be a 'good boy' or a 'good girl', to be validated by others opinions about us. Then at school, we're taught to strive to be top of the class, to pass not fail our exams, to get the highest grades we can. Why do you think the 'like' button on Facebook became such a popular and widely copied feature? In short, we're taught to seek the approval of others, and taught that failure is a thing that will bring the disapproval of others, a thing to fear, a thing that must be avoided at all costs.

You can understand why in a way. Parents and teachers only want the best for the children in their care, and children don't know the best ways to grow and develop in order to survive and thrive in the world, so they do need guidance. However, somewhere along the line, our care for them has gone from being interested in their growth and development, to caring specifically about their success against an externally approved set of criteria, against which they will be judged as passing or failing.

The problem is that this ignores the fact that failure itself is a form of growth and development. In late 2017 I became a parent for the first time, and during the time I've been writing this book, my daughter has been learning to walk. Frankly, to begin with, she was rubbish at it. Constantly falling over, tripping over things and only managing a step or two before having to sit down again with a bump. The thing is, this constant failure didn't bother her at all. Unless she'd fallen and actually hurt herself (which she rarely did), then every time she fell down, she just got back up and tried again. Doubtless she was learning lessons from each time she fell down, and was using these lessons to help her not to fall down again. Essentially, she kept on failing, and through failing, she learned to be a success at walking.

It was fascinating watching her do this, as it struck me just how different her approach to growth and development was to the thousands of adults I've worked with throughout my career. So often in the workplace, people have learned to fear any sort of failure so much that they go to extreme lengths to cover it up, or put a positive gloss on it, or just claim it never even happened at all. Yet what a shame this is, because I'm sure they too were once small children who learned so much through repeated failure, just like my daughter did.

Now I imagine that by this point you've got the same funny look on your face as some of the people on my training courses have when I raise these points. Whilst what I've said may make logical sense, surely I can't mean that we should be encouraging people to fail at work? Failure's not something we can welcome universally. Some failures are risky, dangerous and destructive. Failure can destroy projects, smash budgets, bring down entire organisations and even kill people. Surely we need to prioritise reducing failure as much as possible, rather than encouraging it?

This is a tempting view, but it's actually mistaken. The reality is that when people or organisations have a fear of failure, they create a culture that's toxic to the idea of transparency. You see, failure is always going to happen. The only thing we can expect about the unexpected is that it's going to occur, in ways that we didn't expect. Things going wrong, or at least not working out the way we expected them to, is to some degree just a natural part of human existence. So if we make failure unacceptable in a world where it's always going to occur, then what we create is a massive incentive to hide and cover up our failures, reducing our transparency.

This is a problem, because transparency is a key cornerstone that both lean and agile approaches need in order to work. In order to continuously inspect and adapt the work we do and the way we do it, we need to be transparent and honest about what is happening. Penalising failure makes this much less likely.

Take for example an organisation that has set targets for reducing the number of defects in the software it is creating. Its aim is to get to zero defects as quickly as possible. Sounds like a worthy aim right? In reality though, whilst setting targets like these may seem tempting as an approach, they are very often counter productive. As we've already seen, the likelihood is that defects will always be at risk of happening, for reasons that just cannot be foreseen at the time.

So if getting the defect count down to zero is likely to be impossible, but the target is to get it to zero, what option do people have but to artificially manipulate the data so that it looks like defects are zero? Rather than make things better, this actually makes things worse, as there may have been things you could have done to reduce the number of defects, but whilst people are being incentivised to cover them up, you've got no chance of identifying what the defect causing problems are and how to fix them.

In reality, what you want is a culture that makes it safe to admit when defects are found, and ensures that a root cause analysis is undertaken, in a no blame way, to help prevent the same issue ever arising again. By encouraging people to admit failure every time it happens, you actually do something to reduce your number of defects.

So what's the got to do with people management? A lot, I would argue. People managers are often instrumental in setting the culture around failure. From the performance management system that holds people accountable at the end of each year, to the policies and procedures that kick in when people do fail, people managers and HR professionals are instrumental in shaping how the organisation approaches failure. For example, in a traditional organisation, someone who regularly admits defects in their code may be disciplined, put on a performance management process or even have their employment terminated. Conversely, in a lean-agile organisation, they may well be rewarded as the active guardian and exemplar of organisational problem solving that they really are. I believe people managers could do a huge amount to shape this debate.

In addition, returning to the point of failure as learning too, if you think it through a bit, it's clear that creating a culture where failure is learning is right at the heart of an HR department's goals too. For if an HR department has learning and development within its remit, and there is huge value in failure as a form of learning, then creating a culture where people learn from failure is potentially a big missing piece of the organisational learning puzzle more generally.

Now what this looks like in detail I'll leave to you. There will obviously be limits to what kinds of failure are and are not acceptable, and new ways to handle failure when it occurs. I certainly have theories about tools and techniques for bringing this to life, but it is not the job of this book to give you all the answers. After all, if it did, it would be taking away your ability to learn how to do this through failures of your own.

Section 8: Employee Wellbeing And Retention

Idea 44. Identify Weak Failure Signals In Employee Wellbeing

> **Hypothesis**: Employees are human beings, and often they try to hide or cover up personal life issues that affect their work, even when the issues are very significant indeed.
>
> **Experiment**: Start to create a safe space at work, where people can in confidence report the weak failure signals in their own lives, genuinely without fear of escalation or reprisal.

In the world of lean, there exists the idea of weak failure signals. The thinking goes like this. When something fails, you notice it, and you work to fix it. A machine breaking down, an outage or data leakage affecting thousands of customers, somebody getting injured in the workplace, that kind of thing. Hopefully when something like this happens, you spend some time looking into what happened in order to ensure that it doesn't happen again.

The thing is though, these sorts of occurrences are very strong failure signals. Something's gone so catastrophically wrong that you can't fail to notice it. However, how many of these failures were preceded by a number of weak failure signals that no one paid any attention to? The funny noise that started coming from the machine all of a sudden, the person who nearly left their laptop on the train but ran back to get it just in time, the near miss on the factory floor. Or if someone did pay attention to them, they didn't report them, because they thought things would probably get better, or that they'd get in trouble if anyone found out what nearly happened.

The tragedy is that if these near misses, these weak failure signals, had been noticed or reported at the time, the strong failure that followed them could have been avoided. Huge amounts could have been saved, be that of money, organisational reputation or even people's lives.

So what has this got to do with people management?

Well, as covered more in idea 43, people managers should be the ones owning and driving a culture of making it safe to make weak failure signals visible, and have them investigated, without the person reporting them feel like they're at risk of criticism or disciplinary action (the idea of 'shoot the messenger'). As well as this role in making it safe to report weak failure signals, and protecting those that do so, people managers should also be looking at how they spot weak failure signals in areas closer to their area of interest too.

To give an example, some years ago, I worked at a company that hired a great new software developer. On paper, their experience was fantastic, at interview they were superb, and we were super excited to have them join us. Only once they did join us, they really didn't seem that great at all. They'd turn up to work late, hardly talk to us whilst they were there, and they often looked quite ill or even hungover. Increasingly the feeling in the office was that they were a slacker, that they'd duped us by getting the role, and that they had to leave, so eventually they did.

A few years later, I was out at a club night when I bumped into this person again. That night they seemed like a different person, happy, healthy and outgoing, just like the person we'd first interviewed all those years ago. In fact, it turned out that they were now being very successful in a senior technology role at a large corporation. It being a club night, I'd had a few drinks, so I decided to be honest and ask just what they thought went wrong when they worked for us. Simple, they said, between the interview and starting the role, their long-term partner had left them, and their life had completely fallen apart. They'd never told us this, but then we'd never asked, and besides, what would we have done with this potential weak failure signal if they had told us anyway?

The example of this person has stuck with me ever since, and I'm sure we've all had our weak failure signal moments over the years too. Times that we'd wished that someone had known exactly what we were going through, and been able to give us support to stop it escalating into a bigger problem, but instead we've had to keep on wearing the 'mask of professionalism' we feel we need to wear in the workplace, regardless of how much our tears are causing that mask to slip off our face.

Due to their role in the formal sanction processes, and their often process heavy ways, people managers can often not feel like the place people can go to to raise weak failure signals around themselves. Even occupational health departments are ultimately known to be in the pay of, and thus on the side of, the employer rather than the employee. This is such a shame though.

Weak failure signals are exactly what you want to be spotting if you want to stop a small problem that's cheap and easy to fix from escalating into a bigger one that's far more costly and difficult. As well as being at the forefront of creating a culture where people can raise weak failure signals and have them investigated without fear of blame or sanction, people managers should also be championing the ability for people to raise weak failure signals about themselves and their real lives outside of work too. If we don't, organisations just risk repeating the mistake we made all those years ago, of losing a hugely talented person, just because we never thought to ask them what was wrong.

Idea 45. Make Neurodiversity A Competitive Advantage

Hypothesis: Workforce diversity gives an organisation a competitive advantage in a complex, fast changing and customer focused environment. This includes neurodiversity, and small changes to working environments can allow the neurodiverse to deliver huge amounts of added value.
Experiment: Explore the employment options that would make neurodiverse people better able to succeed, from hiring to retention, and implement them as small changes to measure the impact.

This is a difficult idea to cover in a way, as it's a slightly sensitive subject, and it's one where there probably needs to be more empirical research around the topic too. However, it's also a hugely important area to consider around your lean-agile transformation; the topic of neurodiversity.

In some ways, the fact we talk about neurodiversity these days shows just how far we've come as a society. What used to be stigmatised issues called 'mental illnesses' are now seen as the very mixed and wide ranging characteristics of many different people, and it is finally being recognised just how valuable the contributions of these people can be. As such, areas such as depression, Asperger's and autism, OCD and many other different types of neurodiverse behaviour are seen as what they are; differences from what has long been accepted as 'the norm', but not things to reject or shy away from. There are a number of good reasons for this.

The first is that diversity is strength. In a lean and agile context, you're constantly looking to innovate, to come up with new ideas and test them out, in order to gain a competitive edge in a rapidly changing external market environment. As a result, the more neurodiversity you can have amongst your workforce, the more different perspectives you can bring to a problem, and the more likely you are to synthesise something genuinely innovative and new from those perspectives. In this sense, neurodiversity is just a more recently recognised area within the wider field of diversity in the workforce, which has long recognised that bringing together people from different genders, cultures, sexualities or any other area of diversity only increases the strength and customer centricity of an organisation.

The second issue though, and this is one where more research is needed, is that in the lean and agile spaces, there possibly tends to be more neurodiversity present anyway. In a 2009 paper, Alison Hunter examined the prevalence of introverted and Aspergic personality types in the computing industry, and suggested that these types of people are actually more prevalent in this industry than in many others. She noted well-known Aspergers researcher Tony Attwood, who said;

> *"One of the reasons why computers are so appealing is not only that you do not have to talk to or socialise with them, but that they are logical, consistent and not prone to moods.*
> *Thus, they are an ideal interest for the person with Asperger's Syndrome."*
> (Attwood, T. 1998, p94).

Now agile especially has its roots in the computing industry. Indeed the original, official title for the 'founding document' for the movement, the agile manifesto, is actually 'The Manifesto For Agile Software Development'. As a result, agile is often applied in an industry that likely has a higher degree of neurodiversity within it, and to take the other perspective, has likely also been influenced as a movement by many neurodiverse people over the years as well. My personal experience is not the same as empirical evidence of course, but in my 15+ years of working in agile and software development, I have certainly come across large numbers of neurodiverse people in these fields, more than I have when meeting people working in other fields.

So neurodiversity is both an asset to an organisation that wishes to be agile, and likely also an inevitable reality to an organisation if it is working to some degree in the technology or software development spaces. What then can people managers do to support this?

Well thankfully, agile actually provides us with many of the answers. The primary one lies in the ideas of self-organisation and self-management. A diversity of personality, thinking and behaviour types within an organisation will require a diversity of approaches in order to accommodate these people.

For example, some people enjoy the social experience of working around other people, whilst others find it overwhelming from a sensory perspective, and anxiety provoking from a social perspective too. Thus we should provide people choices around their workspace, allowing them the freedom to choose to work in groups if they wish, or alone if they find that easier. For some people, coming into an office environment can be stressful and distracting, and they in fact work better from home. I'm certainly one of these people, and find I work far better from home as my default option, with occasional trips into an office for meetings if required.

Some people find change very difficult, even down to the tools and devices they are asked to use in the workplace. Personally, I'm an Apple Mac user, and I find it extremely stressful to have to use a Windows based PC. Other people are the opposite. An organisation that seeks to benefit from neurodiversity should let people choose their own tools, even bring their own devices to work.

When it comes to hiring for neurodiversity, you may need to think of alternative approaches too. For example, I was once helping out with an organisation's graduate recruitment programme, where they had an interesting challenge. Because they knew that graduates wouldn't yet have the skills and experiences a role would normally require, they'd developed a system for hiring on strengths. For example, did the person have good communication skills? What were their interpersonal skills like?

The only problem was that whilst this approach may work well for hiring graduates into a fast-track management scheme, where communication and interpersonal skills were likely to be essential, it proved very difficult when hiring people into technology jobs. Some of the most promising coders and software developers had chosen that field for exactly the reasons that Tony Attwood described above, because being brilliant with computers means you don't have to be brilliant at communication and inter-personal teamwork.

Even if some of these people did make it through the hiring process, apparently the graduate programme itself sometimes accidentally forced them out, as it was designed to cycle its participants rapidly through a series of different role placements over a couple of years. These were people that found change extremely difficult to cope with, so what would normally have been an exciting two years of ever changing experiences and opportunities for a neurotypical graduate became a hellish two years for a neurodiverse one.

Assuming these people do manage to join an organisation and stay in it for a decent length of time though, the chances are that the usual promotion mechanisms are still going to fail them. For example, does your capability framework unwittingly have neurotypical behaviours written into it, such as communicating well with others, having widespread networks of influence and so forth? What about the unwritten side of the promotion game? Going for drinks with the right people after work, networking well at industry events, even something as simple as being seen to be joining in at the office Christmas party?

These are all, to a great or lesser degree, not difficult behaviours for neurotypical people, but they are potentially impossible ones for the neurodiverse. Incidentally, one of the kindest things I once saw a organisation do for neurodiverse people was rather than just funding an office Christmas party, the organisation said that people could attend the party if they wished, but if they didn't want or feel able to do so, they could claim the same amount of money their party attendance would have cost on expenses for food and drink of their choice. One neurodiverse person I knew there used the money to go out to a restaurant by themselves. They took a book, ate some great food, drank some nice wine, and had a really lovely time as a result.

So the answer to gaining a competitive advantage through embracing neurodiversity lies in providing a diversity of options for the neurodiverse. Options that are as diverse as the diversities that exist within your organisation. Perhaps through its roots in a community that is itself more likely to be neurodiverse, agile provides an excellent answer to these challenges. Give people a diversity of options around their different working styles and practices, then let them self-organise to find the ones that suit them the best.

Idea 46. Do You Need A Sustainable Pace Or Just A Regular Servicing?

> **Hypothesis**: Some people find working at a sustainable pace extremely difficult, but they are still at risk of burnout which much be prevented.
>
> **Experiment**: Put in place systems to support people who are unable to work at a sustainable pace, to ensure they do not burn out.

Last summer I had a neighbour round for a barbecue. He lives and works locally, and we got to talking about our respective commutes to work. I mentioned that with all of my international travel last year, I'd done 67,000 miles of travel, or about two and a half times round the world, and it blew his mind, as he only works just down the road. Yet I've never seen this as something to be jealous of. You see, I travelled so much because I threw myself into my work 110%. I exhausted myself, missed the first year of my first daughter's life and the quality of my work at times dropped, alongside all sorts of other similar negatives things.

It's for this reason that the agile manifesto talks about the importance of maintaining a sustainable pace. For if people don't work at a sustainable pace, they get tired, they burn out, and the quality of their work drops. When the quality of their work drops, the cost of rework and error fixing increases, and the pace of delivery slows down.

However, I've personally always struggled with the way the agile manifesto expresses this. For it says;

"Agile processes promote sustainable development. The sponsors, developers, and users should be able to maintain a constant pace indefinitely." [13]

Yet this is never how I've worked. When I'm working on something I really care about, I work far too hard at it, working at a pace I can only really maintain for three to six months, sometimes less. I don't think I'm alone either. Whether you call it flow state working as discussed in idea 19, or passion for your work, or even obsessional interests found in neurodiverse conditions such as autism, there are many people out there who work 'too hard' then burnout from it. Asking them to work at a sustainable pace is just alien to their nature. They're all or nothing people. But how do we deal with them?

Well, let's use cars as an analogy. Your car doesn't run at a sustainable pace, constantly moving at a moderate speed for eight hours per day. Sometimes it just drives to the local shops. Sometimes it drives at high speed for hundreds of miles in one go. Sometimes you park it up and it sits there doing nothing for a while. But how does your car deal with this lack of constant and sustainable pace? Well, you take it for a servicing. In contexts where things may be working at unpredictable rates and are at risk of burnout, you stop regularly to look at them, check they're ok, do any work on them that needs to be done, then get them back to work again.

It's this sort of idea I'd like to propose for people like me who are unable to work at a sustainable pace. In fact, it's a way of working I've successfully implemented during my career so far, sometimes more by accident than design. Every few years I take a few months off work to get a complete break from it all. Sometimes this has been caused by redundancy, sometimes by me resigning from a job, sometimes thanks to parental leave after the arrival of a new child. But each time I've gone back to work better, fitter, happier, stronger and producing my best

---

[13] http://www.agilemanifesto.org/principles.html

work once again.

Now, how great would it be if provision could be made formally in the workplace for people who don't work at a sustainable pace to enable them to stop, recharge and produce their best work again? As discussed in idea 7 in this book, this would be the sort of thing that would be difficult to do in a workplace that only measures hours worked. But in one that had switched to measuring value produced, it could well be feasible. People need to produce X amount of value each year, and it's up to them and their working style whether they do that at a steady pace, or whether they work too hard, then stop, then work too hard again, etc. The act of stopping could be seen as the equivalent of taking your car in for a check too. People could have a check in with their manager to see how they're doing, perhaps a check in with occupational health to see how their health is holding up under the strain, that kind of thing.

You could of course argue that letting people stop work for a while and take a break is unfair on those people that do work at a sustainable pace. There are two answers to this. The first is perhaps the easiest. If you have employment contracts that say people only need to work, say, 35 hours per week, then let them be militant about their timesheets and in return, be generous with your 'time off in lieu' policy. For example, when I run training courses, I generally work a 14 hour day at least, as we're in the venue setting up way before the attendees arrive, then we spend hours in the evening debriefing the day and planning how to make the next day even better. So if I've run two four-day courses, I've effectively worked a whole extra week. Store those weeks up over a year or two, and I should be able to take a decent break.

An even better option of course would be to stop worrying about how many hours people work at all, and instead measure the value they deliver. If they deliver all of their objectives for the year by August, then that should be them done with work for the year. I admit that might be a leap too far for many organisations however.

There is though one other problem with all of this. The implicit undercurrent of servicing a car is that if it's too broken, or uneconomical to fix, you scrap it and get a new one. This works because cars aren't sentient. They don't have any idea what might happen to them when you take them in for a servicing, and they don't have feelings such as fear and upset as well. People, however, do. If someone knows that if they go for a check in before a break, they risk getting sacked if they fail, then they won't go for check ins and the whole system breaks down. Again, the importance of psychological safety in the workplace shines through.

So if you're going to run a system like this, you first of all need people who can only work at an unsustainable pace, or at least who primarily prefer to do so. I'd wager you do already have people like that, we're pretty common. Then you then need a system for them to be able to stop for a bit, get a check up, and rebuild their strength before starting again. You also need it to be a system that doesn't put people at risk of sanction or dismissal if they are discovered to have developed problems from their unsustainable pace. Ideally you need a system that measures people on value delivery rather than hours worked too.

This may all seem like a lot to put in place, but I suspect the rewards from it are huge. Some people will always be working at an unsustainable pace, no matter how much the agile manifesto tells them not to do so. At the moment they're burning out, going off sick or leaving the organisation, which are all significant forms of waste. How much better would it be if systems could be put in place to recognise and support them, given the huge value they often create when they are working at pace?

Section 9: Leaving A Role

Idea 47. Stop Accidentally Sacking Your Agile Talent

> **Hypothesis**: You may be spending more time attracting lean-agile talent than you are ensuring that it doesn't leave.
> **Experiment**: Set up regular, informal sessions with identified talent to hear their frustrations and concerns, to pick up early any hints that they might be looking to leave.

It's a tough market out there for organisations looking to benefit from lean and agile approaches. You only have to look at how many long term traditional project managers are rebranding themselves as 'Enterprise Agile Coaches' on Linkedin to see that there's clearly more demand than supply right now. So with all of this noise in the candidate market place, finding real agile talent, people who genuinely understand, live and breathe what agile means, can be really tricky.

We talk about how to find and hire these people in section 4. However, for some organisations there may be no point in putting huge time and effort into hiring these people. For no matter how good you might be at hiring them, your organisation may accidentally be firing these people on a daily basis. Here are four common reasons why this might be happening.

The first is transparency. Agile people love transparency. It's right there in the core principles of the Scrum framework, and the Scaled Agile Framework (SAFe), and the Large Scale Scrum (LeSS) framework, and in most other agile approaches you'd care to look at. The problem is, transparency is very often an alien concept in the non-lean-agile enterprise. In a world where knowledge is power, and appearances matter more than realities, transparency is something that doesn't win you friends and influence people.

One very common occurrence is that a new agile person turns up in an organisation, starts creating transparency, and suddenly everything seems far worse than it did before. Things aren't worse of course, but the agile person has turned the problems that were previously hidden into issues that are now visible for all to see. That's why transparency is written into these agile frameworks, so you can see all your problems more accurately and start to fix them.

However, in a world where appearances matter, what's the easiest way to fix the appearance of the problems newly brought to the surface by the arrival of this agile person? Get rid of the agile person of course. In reality, their sacking won't make anything better, just as their arrival didn't make anything worse, but at least everyone can go back to sweeping problems under the carpet and everything will look fine again. Shoot the messenger, the reasoning goes, and hopefully the uncomfortable message will go away. You particularly see this where managers have got themselves into a position of seniority by being seen to be standing next to the successes and standing as far away from the failures as possible. The moment someone turns up and starts highlighting failures in their area is the moment they decide to dispense with that person's services, and any hope you had for adopting agile approaches disappears with them.

The second reason is frustration. For this, let me share an example from history that exemplifies well how lean-agile people often feel. In 1990, one of the things that ultimately led to the demise of influential British Prime Minister Margaret Thatcher was the resignation speech of government minister Geoffrey (now Lord) Howe, who felt he could no longer continue to serve under her as part of her government. Much of his resignation speech related to the current situation of the UK Government at the time, but some of its most influential passages illustrate perfectly how agile people often feel in traditional organisations.

In his resignation speech, he said;

"(serving under Margaret Thatcher) is rather like sending your opening batsmen to the crease only for them to find, the moment the first balls are bowled, that their bats have been broken before the game by the team captain."

Too often, great agile talent feels just like this. They try to do what they can, they try to build great teams, coach them in new ways of working, and then they suddenly find that someone more senior comes in and undermines everything they are trying to achieve. Sometimes agile people just get up and leave as soon as this happens. After all, when agile contract work is plentiful, some people just aren't looking for situations where their agile practices will be continuously undermined.

Agile people who resign often take others with them too. I've lost count of the number of friends who have resigned from places, and have then tried to persuade me to jump ship too and go and join them in their new organisation. This sort of sentiment is summed up well by this second quotation from Lord Howe's resignation speech;

"The conflict of loyalty, of loyalty to my Right Honourable Friend the Prime Minister…and of loyalty to what I perceive to be the true interests of the nation, has become all too great. I no longer believe it possible to resolve that conflict from within this Government. That is why I have resigned. In doing so, I have done what I believe to be right for my party and my country. The time has come for others to consider their own response to the tragic conflict of loyalties with which I have myself wrestled for perhaps too long."

Why would someone want to stay in a role where there work is constantly being undermined? Why would they not try to take others with them when they leave? Both of these are very common and powerful problems during a lean-agile transformation.

The third reason is invisibility, and this stems from the reason that some of the agile mindset is an odd thing. In most traditional organisations, the way you get ahead is by being seen standing next to the organisation's successes, and standing far away from its failures. However, agile often has a strong theme of servant leadership running through it, a school of thought originating from a man called Robert Greenleaf.

Servant leadership is a really interesting school of thought if you fancy looking into it further, but one of its biggest impacts in an agile context is the fact that its practitioners gain their gratification at work by serving others, helping them be brilliant and successful, rather than wanting to be seen to be brilliant and successful themselves. For example, if they write a report on a piece of work, they may put the name of the person who delivered the work on the report, rather than their own name, even though they wrote the report. They spend their time doing everything they can to help others be great, and have little interest in their own profile within the organisation.

The problem then comes when you come to something like the end of year performance appraisal, where each person is expected to write down all the ways they've personally been successful in the last year. A servant leader may genuinely find it hard to write anything, as all their time was spent helping others to be successful. But if you don't succeed in your appraisal, your time in the organisation may be drawn to an end.

Robert Greenleaf was inspired to create the field of servant leadership by Herman Hesse's book 'Journey To The East'. In that book, a long journey being made to the east is supported by a servant called Leo, who appears to be just the mere servant for the people on the expedition. However, when Leo leaves, the whole expedition falls apart. The reality of servant leadership within organisations can be similar. The person who seemed like they were achieving nothing turned out to be the lynchpin for the whole delivery, but you only find that out when they leave.

The fourth and final reason is one I call 'not being trapped in the game'. More than once in my career, I've found myself disappointed by something at work, only to be told that I'd have done better and been happier if I'd 'played the game'. This usually means doing something I wouldn't naturally do, or wouldn't agree with doing, but that needed to be done in order to win a reward, usually from senior people. There are two aspects to this game that cause problems with retaining agile talent. The first, of course, is that agile people newly joining an organisation will have far less knowledge of what the game involves than the people who have been there for some time, and where people see that game as being 'zero sum', that it to say when one person gains another person loses, then the agile people will be continually be out competed, leading to their demoralisation and desire to leave.

The second aspect though is that the existing people, if they've been there sometime, often find themselves trapped in the game. In other words, they've played the game at one organisation for so long that they no longer know how to go and play a different game somewhere else, or at least they don't have any confidence that they could do so. As a result, the motivations of the existing people and the agile people differ significantly. Whilst the agile people know that they could easily go and find work somewhere else, the existing people will fight tooth and nail to win the game, because their fear of losing is so much greater. As a result, their game playing may force out agile people, who have neither the same experience in the game or the same motivation to win it as the existing people.

I'm sure there are many other patterns that accidentally force lean and agile people out of organisations as well. These are just four of the most common that I've noticed repeatedly over the years. If you've experienced any others, I'd love to hear about them, do please get in touch. However, the most important lesson from all of them is that the accidental firing of talent is a very real phenomenon that you have to be constantly vigilant towards. In a world where genuine lean and agile talent is in short supply, an investment in ensuring that they don't leave could be even more valuable than investing in attracting them in the first place.

Idea 48. Make Talent Retention A Collective Responsibility

Hypothesis: People leave bad organisational systems more than they leave bad managers

Experiment: Start to engage senior leaders with understanding the impacts of the holistic system on talent retention, and with the idea that talent retention should be treated as a collective responsibility.

A while ago, I was catching up over drinks with a friend, a fellow agilist, and we got to talking about how work was going for them. As far as I was concerned, they were one of the best agile coaches that I knew, but it turned out that deep down they had a problem. Being so great, they were regularly being put at the forefront of the change initiatives in their organisation, and as a result, they constantly felt pretty precarious in their role. As they put it, they were only ever one or two 'wrong moves' away from being forced back into a 'delivery role', or effectively dismissed from the company (see idea 30 for more on this common phenomenon).

Whilst we were chatting, we began to realise something. Them staying in their role wasn't due to one single thing or one single person. The cause of their job ending could come from anywhere at any time. As a proactive change agent, they came in contact with numerous different people from across the organisation all the time. Any one of them could either take against them and put a complaint in against them, or 'have a quiet word' as such things are done in the UK. Alternatively, any one of them could annoy my friend so much that they just decide to throw in the towel and quit.

However, despite this realization that the reason for their job ending could come from anywhere at any time, many people interested in talent retention still believe that people don't leave organisations, they leave managers, specifically their direct line manager. So if their manager isn't great, or if they don't have a great relationship with them, then the employee will be more likely to look elsewhere. Now, management is of course a factor in where people choose to work for sure, but there's research out there that shows that other factors are just as big a factor, if not bigger.

For example, there are Dan Pink's ideas that people value having autonomy (the ability to control how they work), mastery (the ability to get better at what they do), along with a sense of purpose (provided by leadership)[14]. Add to that the ideas from systems thinking, that the problems people experience are likely the effects of the whole system within which they work, even if they feel like single, localised issues.

These sorts of ideas are backed up by surveys and empirical research too, which have found that areas such as leadership, along with personal growth and development, are much more influential in people's decisions to stay at an organisation or leave it than something as simple as who their manager is. In many ways, you could argue that seeing someone's immediate manager as the problem is a phenomenon caused by a lack of systems thinking, again confusing a localised symptom of a systemic issue with its wider cause.

So what would be the lean and agile solution to this? Well, if the problems are likely systemic, not under any one person's control, then why should we not apply agile ideas of collective responsibility and continuous learning to them? Just as each member of a team in Scrum is collectively accountable for the delivery of the work of the team, regardless of their job title, why shouldn't the senior management and/or leadership of an organisation be collectively accountable for talent retention? If someone decides to leave an organisation, then shouldn't everyone within the system be asked to look at their potential role within that, rather than just laying 'blame' at the feet of their direct manager, or even worse, just shrugging and hiring someone else?

---

[14] See 'Drive' by Dan Pink in the references section at the end of this book.

On top of this collective accountability should be a collective desire for learning around talent retention. Not just running an exit interview with someone (which they might not bother to attend[15]) and filing the results away somewhere in order to meet the agreed policy, but genuinely finding out what their drivers were for leaving, building a picture over time of whether there are patterns being repeated across different groups of people, or even within specific groups, then using those patterns to identify techniques for ensuring talent retention is increased and the collective goals around it are achieved.

The primary goal of an organisation is delivery of value for a set or sets of people. The senior management and/or leadership of an organisation is accountable for ensuring that value delivery occurs. That value is largely delivered by the people the organisation employs. If those people leave, their potential for creating value for the organisation leaves with them.

Put all of these things together, and it becomes clear than in a lean-agile organisation, focused on delivering value, talent retention should be a high-priority and collective responsibility, shared equally across a senior set of people, who collectively should be relentless in understanding how talent retention works within their organisation, and relentless in learning about how it can be improved.

---

[15] As another friend once commented when they were offered an exit interview, *"Leaving a job shouldn't mean you have extra homework to do"*.

Idea 49. How To Handle Redundancies In Lean-Agile Transformations

> **Hypothesis**: Cost saving through redundancies is often a motivation for a lean-agile transformation, but often this is a counter-productive goal that, especially when handled incorrectly, can significantly harm the success of the transformation itself.
>
> **Experiment**: Start from the premise that your existing people are a hugely valuable resource, and focus on maximising the value they deliver, not the costs they incur.

It's a tricky topic this, but that's all the more reason to talk about it. Although creating redundancies isn't mentioned in any lean or agile framework, it's actually a relatively common occurrence during the transformation process.

Often, this is because the driving factor behind the transformation is actually cost saving. Senior execs have heard that lean and agile are great ways to 'do more with less' and as a result are happy to fund a bunch of consultants to come in and run a transformation, on the understanding that the initial costs will be recouped over time through all the waste that can be cut out of the system. The waste, very often, being seen as being people. Common as it is, this approach is actually very mistaken.

Lean and agile can allow you to do more with less for sure, and I don't doubt that any large organisation will be able to find a number of different types of waste in its system that can be cut out. However, their actual purpose is to enhance value delivery, not just cut out waste. To look at them as waste reduction is to miss their potentially much larger benefit. Rather than thinking that we could get the same amount of value out of our organisation as we do currently, just with fewer people and lower costs, thus increasing profit margins, what if we thought about getting even more value out of the organisation with the same number of people, which might actually increase margins even more?

Taking a 'transformation with redundancies' approach is also very counter-productive for the success of the transformation itself. A successful transformation is going to require people to be bought in at every level, and make some substantial changes to their mental models of how they see the world. This is a difficult enough thing to achieve at the best of times, but when you send the message that the transformation may be threatening your job or the jobs of your friends, you're hardly going to win the battle for hearts and minds that will be required. It is for this reason that Martin and Osterling, in their book 'Value Stream Mapping', say;

> *"If the inspection and improvement work will lead to redundancies, the organisation should make these redundancies before starting on the transformation, so everyone else has psychological safety."*

However, the ultimate reality is that if the organisation is going to make substantial changes, then some roles will become redundant, potentially being replaced with new roles more appropriate to the new organisation that is being created. So what do you do about this?

The first thing to focus on is a core belief in the value of the people you employ already. Just because you're changing the way the organisation used to work, it doesn't mean that the organisation will be 100% different from how it was before. Some elements of organisations change more easily than others, and some may never change at all. So the people you employ currently hold something that will continue to be extremely valuable; the knowledge of how the organisation really works. This is made up of many elements, but can include knowing what approaches have been tried and have failed before, or knowing the right people you need to engage to get things done, or more generally understanding the 'shadow system', that intangible mix of people and culture that is the true reality of how the organisation actually works, regardless of what its policies and procedures say. In short, your current employees have a huge amount of valuable knowledge in their heads. To see them as the 'old guard' and throw them out in favour of new people is a recipe for failure.

However, as mentioned above, these people will need to learn new skills and work in different ways. There are probably two key elements in enabling this. The first is in giving people the freedom to choose what their new role might be. In an organisation that's used to a command and control approach, where people are seen as 'resources' to be deployed in whatever way senior managers require them to be, then the temptation might be to reassign people based on some new operating model designed by senior management. I believe this would be a mistake. There is no one-to-one mapping between traditional roles and lean-agile roles. Things that used to be done by one person may now collectively be owned by a group of people, and vice-versa. As a result, you can't really map things out neatly, and so as in any complex situation, it's likely to be far better to devolve that decision making down to where the information is, by allowing people to volunteer for whatever new role they think would suit them best.

The second element is in providing unlimited and extensive training that anyone is welcome to attend. You can't expect people to self-organise into and volunteer for new roles if they don't understand what those roles are in the first place. You can't expect them to be great at their new role unless they have the chance to learn as much as they can about how to be great at it. Most organisations adopting lean and agile understand that re-training will be required, but again the temptation for traditionally minded organisations will be to try to map this re-training out, and decide who should be assigned to which course. In reality, a far stronger approach, and one that clearly demonstrates that the organisation understands the lean-agile world it is trying to move into, is to provide open and unlimited training to all, and allow people to self-organise around what training to take and when. How you might build a training system that copes with this is covered in idea number 34.

However, once you've tried both of these approaches, there may still be some people left for whom this whole lean-agile thing is just not for them. That's fine, that's natural, and it has to be acknowledged. For if you ignore it, you risk creating people within the organisation who are left with no choice but to say they support the new ways of working, whilst all the time internally hating the new world. Like an undercover resistance movement, these people have a strong motivation to act in a clandestine manner to bring the transformation down from within, in order to return to the old world in which they were happy. This can be fatal to the longer-term success of your transformation. For more on this idea, see idea 50

The thing to do with these people is twofold really. First, if your organisation is big enough, provide them a route out into a different part of the organisation where their more traditional skills are still valuable. Lean and agile aren't solutions for every problem in the world, and it is likely that more traditional skills will still come in handy elsewhere. Again, why throw away organisational skills and knowledge when you could repurpose them more usefully elsewhere. If that isn't an option though, then maybe offering voluntary redundancies is. However, these must be voluntary, and I'd argue should go hand in hand with demonstrating a care for these people even as they leave. Give them access to the training that everyone else has access to whilst they're still there, even provide them coaching and support to find new roles outside of the organisation, and hang on to them until they find something. How you treat those that are leaving will be seen by those that are staying, and will have a big impact on how your transformation will continue to be perceived internally.

Idea 50. Go Fire Yourself

> **Hypothesis**: People may identify themselves or their roles as waste within an organisation, but they have a huge incentive never to acknowledge this or let it be discovered.
>
> **Experiment**: Give people the psychological safety to admit when some or all of their current role is waste, and a safe and blame free way of finding new work that still suits their skillset and personal goals.

I've had an odd career at times in my life, not least because sometimes I've decided to sack myself from a job. That sounds like an odd thing to say but it's true.

Most people, when they leave a job, do so because they're moving on to a new and better job, or returning to education, or taking some time out to go travelling or raise a family. I've done those things too, but I've also sometimes resigned from roles because I believed my role was redundant, or that I wasn't the right person for the role, regardless of what my employer at the time thought. It's led to some interesting conversations over time, along the lines of;

> Them: *"You're resigning?"*
> Me: *"That's right."*
> Them: *"Where are you going to go next?"*
> Me: *"Well, back home I guess. It's where I keep all my things."*

The thing that's caused me to do this is my firm belief in lean and agile as ways of living and working. As a result, I am constantly on the lookout for waste within an organisation, and am looking to remove it. So if during my looking around, I identify myself as waste, then it's myself I must remove from the organisation, even if my employer hasn't yet spotted that I'm waste. You can't talk about lean and agile values and not live them yourself.

The thing is, as odd as I may seem, I actually think there's something important here for lean and agile organisations, especially ones that are undergoing transformations. You see, if you believe in empowering people in their roles, and decentralising decision making, then logically that should extend to them assessing their own performance for the role, and deciding whether they are the best person to carry on in the role or not. I suspect many of us have sometimes had jobs that we've not felt well suited for, and have known deep down that we would be better off leaving. But what support and help is there available for people who feel like this? Probably none.

That's a real shame, as if people had the ability to hold their hands up and say they weren't right for the role, but they were still good, motivated and talented people, then surely there should be the support and psychological safety available for them to hold their hands up and move roles, without feeling at risk of losing the income that feeds their families? Even if they still feel that they're right for the role, do they feel safe to flag up that even part of their role is redundant? Could they do so without the fear that they will then have their whole role made redundant, or that they'll have new things added to their role that they don't want to do? Are we unwittingly creating a situation where people are effectively incentivised to carry out pleasant but wasteful work, for fear they will be forced to take less pleasant but more valuable work if they speak up about the waste?

On top of this, setting up a system where people could be supported to self-identify their role, or even just aspects of their role, as waste could massively help with the wider organisational shift to lean and agile as well. Think of it this way. The current organisation is probably quite happy in its state of equilibrium, and the people are quite happy in their roles. One day, some new people turn up and start saying that the world is going to change, that a new order is going to be imposed, and that the old ways must be removed. How do the people in the old order feel at that point? Like they're being attacked by an invading army? Often times, I'd say they do.

Now here's the thing about invading armies. What do they want to happen to their 'enemy'? Your first thought is probably that they want them to be killed, but actually that's not what invading armies want to happen at all, unless something's gone very wrong indeed. There are actually three options that could happen to the 'enemy'.

The first is be killed, but then you win the war but lose the peace, as who's left to rebuild the country or organisation once the war is over? No one, all you've done is conquered a barren wasteland.

The second option is for the enemy to go underground and let the invading army 'win', but then bring the victors down through a slow battle of resistance and guerrilla warfare. This is actually what happens in many lean-agile transformations. The people being 'transformed' all smile and agree with the new way of working, then sit back and make things as difficult as possible for it, until the pressure gets taken off, the trainers and consultants go away, and things return back to how they were before.

The third option is for the enemy to surrender genuinely, and this is the option that invading armies generally want to happen. If the enemy stop fighting, then no-one on either side is at risk of being killed any more, and you can get on with rebuilding things in the new order.

It is this third option that is so often absent from lean-agile transformations, and one that people managers could step in and provide. For example, take a manager who has been in their job for 20 years, and doesn't feel that they can do anything else. What are they going to do when the new way or working arrives? They're going to enthusiastically support it, then do everything they can to redefine it to mean the status quo, usually in the name of 'being pragmatic', in order to keep their role just as it was. What other option do they have?

Well, people managers could work alongside the transformation team to give people just this option. To say your role doesn't exist in its current form in the new world, but don't worry. We still think you're a great person, with great skills and tons of organisational knowledge. So we can offer you either as much training and support as you want to make the change into the new world, or we can help you find a new role somewhere else in the organisation where you can carry on using the current approach and skill set that you have. Basically, you give people somewhere to surrender to. If you want more thoughts on this, you can also see idea 49.

This approach can then be expanded and maintained throughout the future of the organisation, so that people know that when they identify waste in their current role, they will be supported to remove it and find something else they'd enjoy and be good at doing, rather than feeling like they can't tell anyone for fear of being made redundant or being repurposed into a role they hate.

I may be an oddity in happily resigning from roles with no new job to go to, just because I identify that I'm waste, but the only odd thing I'm doing is doing this without any support from elsewhere. The point of me doing it is still entirely correct, and I suspect both individuals and organisations would benefit significantly if the support to follow my example were provided within organisations as standard.

References

Anderson, D. (2010). *Kanban – Successful Evolutionary Change for your Technology Business*. Blue Hole Press.

Attwood, T. (1998). *Asperger's Syndrome: A guide for parents and professionals*. London: Jessica Kingsley Publishers Ltd. cited in Hunter, A; (2009) *High-tech Rascality: Asperger's Syndrome, Hackers, Geeks, and Personality Types in the ICT Industry*. New Zealand Sociology, Volume 24, Number 2, 2009

Beck, K. (2005). *Extreme Programming Explained*. Upper Saddle River, NJ: Pearson Education Inc.

Deming, W.E. (2000). *Out Of The Crisis*. Cambridge, Massachusetts: The MIT Press.

Deming, W.E. (1993). *Notes from a meeting by Mike Stoecklein*. Taken from https://blognew.deming.org/2015/02/a-bad-system-will-beat-a-good-person-every-time/

Kahn, William A. (1990). *Psychological Conditions of Personal Engagement and Disengagement at Work*. Academy of Management Journal. 33 (4): 692–724.

Liker, J. (2004). *The Toyota Way: 14 Management Principles from the World's Greatest Manufacturer*. New York, New York: McGraw-Hill Education.

Martin, K. and Osterling, M. (2014) Value Stream Mapping: How to Visualize Work and Align Leadership for Organizational Transformation. New York, New York: McGraw-Hill Education.

Oosterwal, D. (2010). *The Lean Machine*. New York, New York: Amacom

Pink, D. (2011). *Drive: The Surprising Truth About What Motivates Us.* Edinburgh, Edinburgh: Canongate Books Ltd.

Senge, P.E. (2006). *The Fifth Discipline.* London: Random House.

Taylor, F.W. (1911). *The Principles Of Scientific Management.* New York: Harper & Brothers